Holt McDougal Mathematics

Course 2
Chapter 10 Resource Book

HOLT McDOUGAL
a division of Houghton Mifflin Harcourt

Printed in the United States of America

ISBN 13: 978-0-55-400718-2
ISBN 10: 0-55-400718-5

1 2 3 4 5 170 12 11 10 09 08

Contents

Description of Contents

Family Involvement Pages

The Chapter Resource Book includes a set of family involvement pages for each section. The family involvement pages consist of the following items:

- **Family Letter**, a two-page letter describing the math that the student will study in each section. A list of vocabulary words from the lessons is included, and some worked-out examples are provided.
- **At-Home Practice**, a one-page worksheet with problems drawn from the content described in the Family Letter. Answers are provided on the page so that the student and his or her family can check this work at home.
- **Family Fun**, a one-page activity sheet that the student and his or her family can work on together.

Practice A, B, and C

There are three practice worksheets for every lesson. All of these reinforce the content of the lesson. Practice B is shown in the Teacher's Edition and is appropriate for the on-level student. It is also available as a workbook (Homework & Practice Workbook)

Practice A is easier than Practice B but still practices the content of the lesson. Practice C is more challenging than Practice B.

Review for Mastery

The Review for Mastery worksheet (one per lesson) provides an alternate way to teach or review the main concepts of the lesson. This worksheet is one or two pages long and is shown in the Teacher's Edition.

Challenge

The Challenge worksheet (one per lesson) enhances critical thinking skills and extends the lesson. This worksheet is shown in the Teacher's Edition.

Problem Solving

The Problem Solving worksheet (one per lesson) provides practice in problem solving and opportunities for real-world applications and for interdisciplinary connections. There are both multiple choice and short response problems. This worksheet is shown in the Teacher's Edition.

Reading Strategies

The Reading Strategies worksheet (one per lesson) provides tools to help the student master math vocabulary or symbols.

Puzzles, Twisters & Teasers

The Puzzles, Twisters & Teasers worksheet (one per lesson) provides fun practice while reinforcing the content of the lesson.

CHAPTER 10

Family Letter

10A Volume

Dear Family,

The student will learn how to identify three-dimensional figures. He or she will learn the differences between prisms and pyramids, and between cones, cylinders, and spheres.

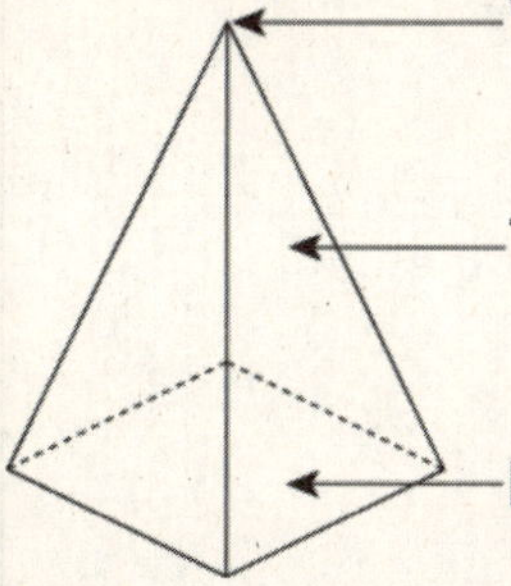

A square pyramid is shown above. It has a square base and all its faces are triangles that meet at a common vertex.

Identify the bases and faces of each figure. Then name the figure.

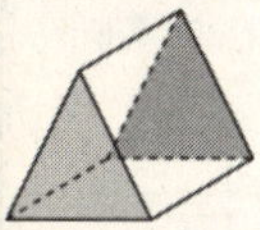

There are two bases and they are both triangles. The other faces are parallelograms.

This is a triangular **prism**.

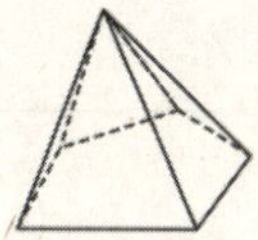

There is one base, and it is a pentagon. The other faces are triangles.

This is a pentagonal **pyramid**.

Some other three-dimensional figures are cylinders, cones, and spheres. These figures are not polyhedrons because their faces are not polygons.

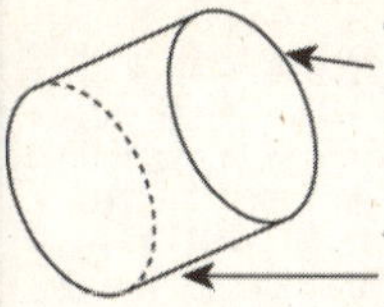

This is a **cylinder**. It has two circular bases.

The curved surface is the lateral surface.

Vocabulary

These are the math words we are learning:

base face used to name a polyhedron

cone a three-dimensional figure with one vertex and one circular base

cylinder has two parallel, congruent circular bases connected by a curved lateral surface

edge a place where two faces meet on a three-dimensional figure

face a flat surface of a three-dimensional figure

polyhedron a three-dimensional figure whose faces are all polygons

prism a polyhedron that has two parallel, congruent, polygon-shaped bases; the other faces are parallelograms

pyramid a polyhedron with one polygonal base and triangular faces that all meet at a common vertex

sphere a three-dimensional figure that has a surface made up of all the points that are the same distance from a given point

vertex of a polyhedron a point where three or more edges meet

Family Letter

10A Volume continued

This is a **cone.** It has a circular base.

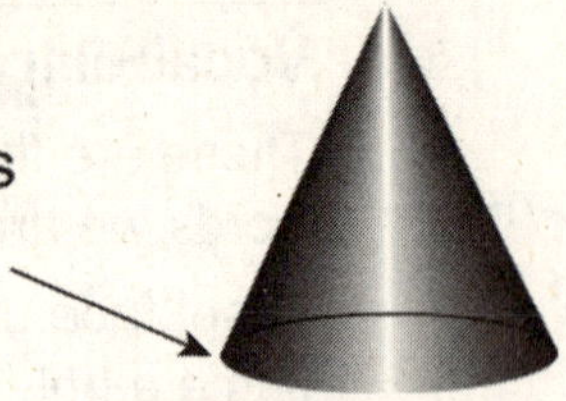

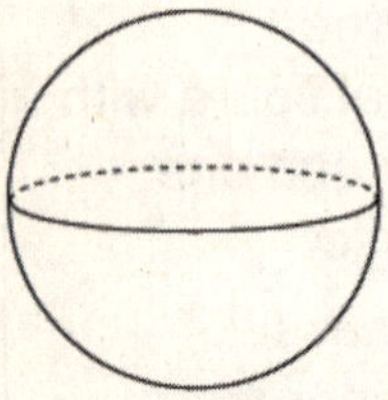

A **sphere's** surface is made up of all the points that are the same distance from the center of the sphere.

The student will also learn to find the volume of prisms, cylinders, pyramids, and cones. The **volume** of an object measures how much the object can hold.

Figure	Words	Volume Formulas
Prism	the area of the Base times the height	$V = Bh$
Cylinder	the area of the Base times the height	$V = Bh$ or $V = \pi r^2 h$
Pyramid	one-third the area of the Base times the height	$V = \frac{1}{3}Bh$
Cone	one-third the area of the circular Base times the height	$V = \frac{1}{3}Bh$ or $V = \frac{1}{3}\pi r^2 h$

Help the student memorize these important formulas.

Sincerely,

volume the number of cubic units needed to fill a given space

Name ______________________ Date ______________ Class ______________

CHAPTER 10

At-Home Practice

10A Volume

Draw the top, front, and side views of the figure below.

1. ______________ 2. ______________ 3. ______________

Identify the bases and faces of each figure. Then name the figure.

4. 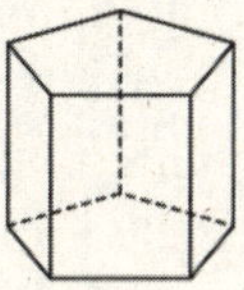

5. 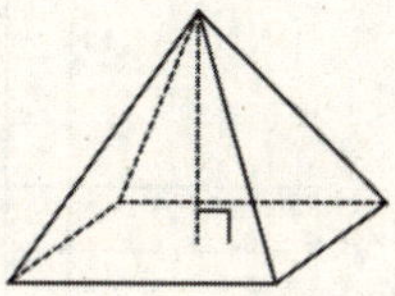

6. 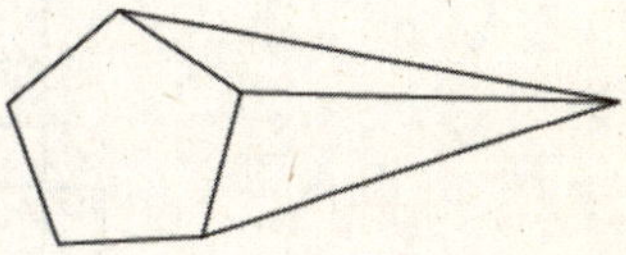

______________ ______________ ______________

Find the volume of the prism or cylinder to the nearest tenth. Use 3.14 for π.

7. 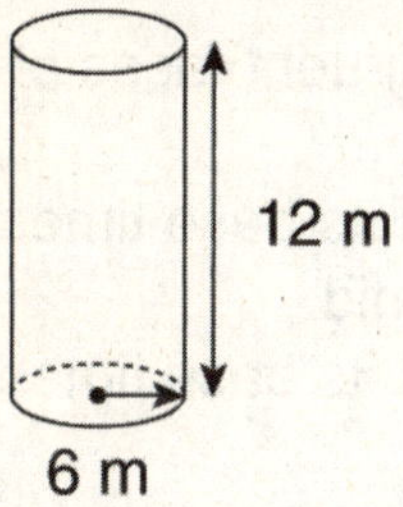

8. 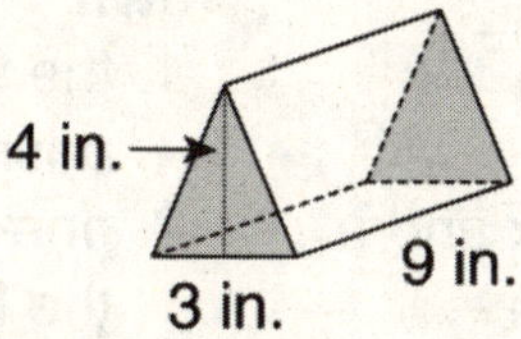

9. 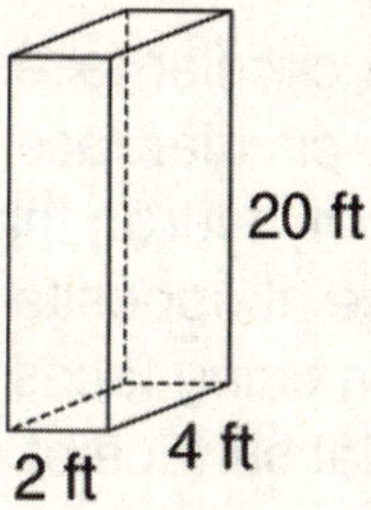

______________ ______________ ______________

Find the volume of each figure. Use 3.14 for π.

10. 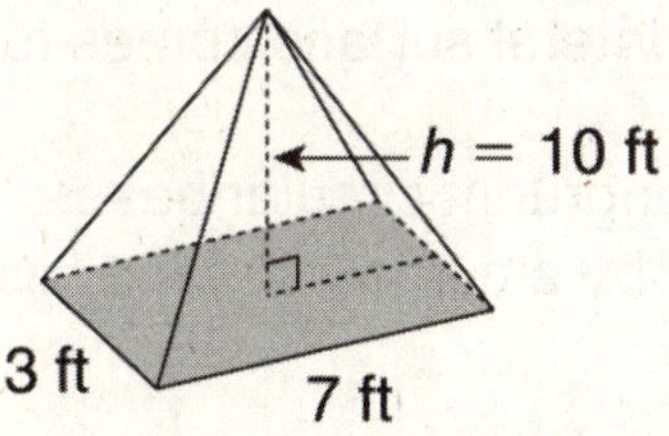

11. 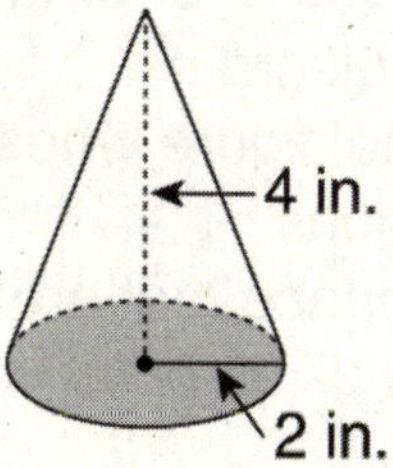

______________ ______________

Answers: 1. 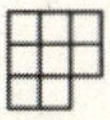 top **2.**  front **3.** 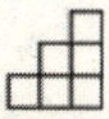 side **4.** 2 pentagonal bases; 5 rectangular faces; pentagonal prism **5.** 1 rectangular base; 4 triangular faces; rectangular pyramid **6.** 1 pentagonal base; 5 triangular faces; pentagonal pyramid **7.** 1,356.5 m^3 **8.** 54 in^3 **9.** 160 ft^3 **10.** 70 ft^3 **11.** 16.7 in^3

Name ______________________ Date ______________ Class ______________

CHAPTER 10

Family Fun

Puzzle

Across

5. the circular faces of a cylinder
7. the circular face of a cone
8. a polyhedron that has a vertex and a face at opposite ends, with the remaining faces being triangles
10. a flat surface of a polyhedron
11. the number of cubes a three-dimensional object can hold
12. a polyhedron that has two parallel congruent bases, with the remaining faces being parallelograms
13. a three-dimensional figure whose faces are all polygons
14. a set of points in space that are equidistant from a given point called the center

Down

1. the two parallel congruent faces of a prism
2. one-third the area of its base times the height of a pyramid
3. one-third the area of its base times the height of a cone
4. has one circular base and a curved lateral surface
5. the face opposite the vertex of a pyramid
6. where the lateral surface comes to a point
9. has two congruent circular bases connected by a curved surface

Answers: Across 5. bases of a cylinder **7.** base of a cone **8.** pyramid **10.** face **11.** volume **12.** prism **13.** polyhedron **14.** sphere **Down 1.** bases of a prism **2.** volume of a pyramid **3.** volume of a cone **4.** cone **5.** base of a pyramid **6.** vertex of a cone **9.** cylinder

Name ________________ Date ________________ Class ________________

LESSON 10-1

Practice A

Introduction to Three-Dimensional Figures

Identify the base of each prism or pyramid. Then choose the name of the prism or pyramid from the box.

rectangular prism	square pyramid	triangular prism	pentagonal prism
square prism	triangular pyramid	hexagonal prism	rectangular pyramid
hexagonal pyramid	pentagonal pyramid		octagonal prism

1. 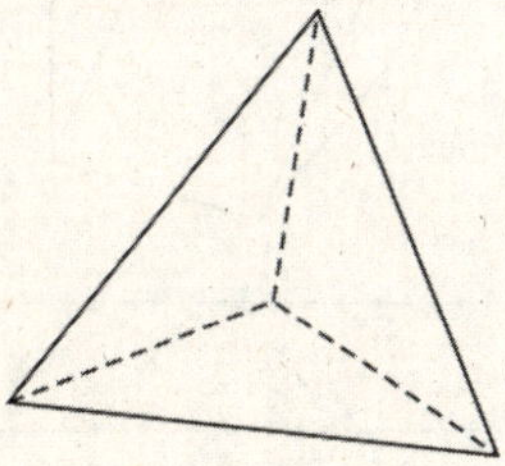

____

____

2. 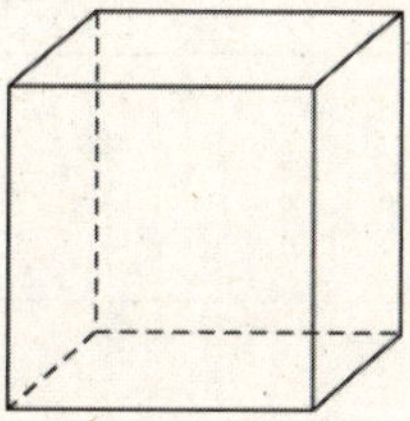

____

____

3. 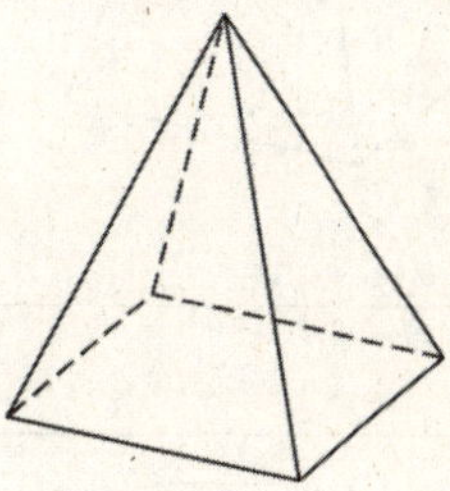

____

____

4. 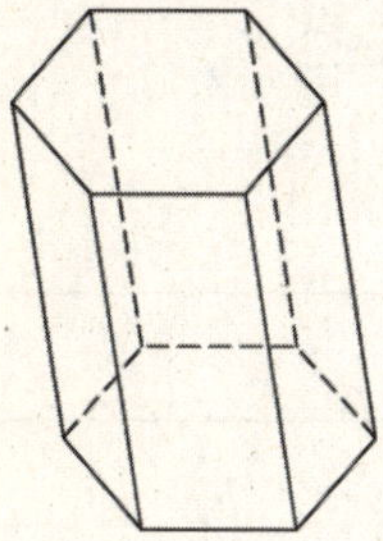

____

____

5. 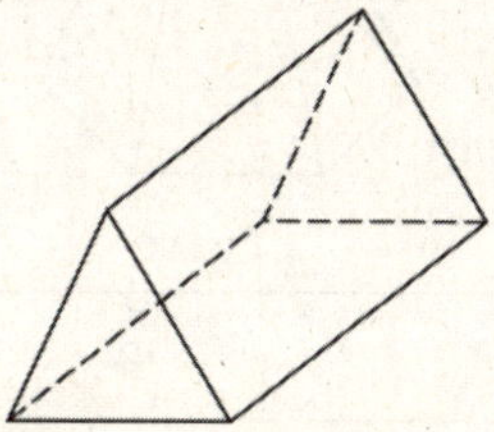

____

____

6. 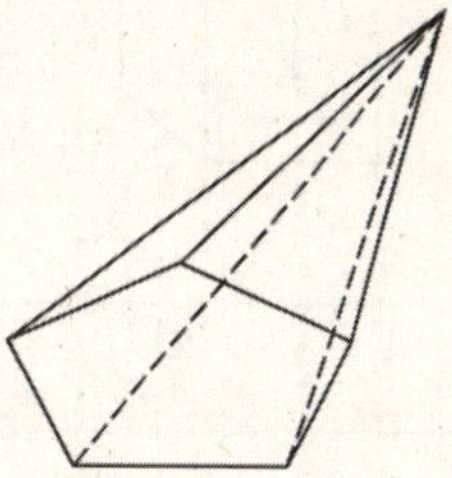

____

____

Classify each figure as a polyhedron or not a polyhedron. Then name the figure.

7. 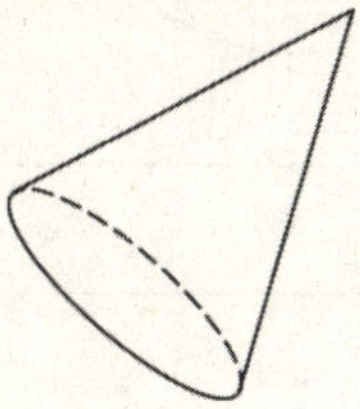

____

____

8. 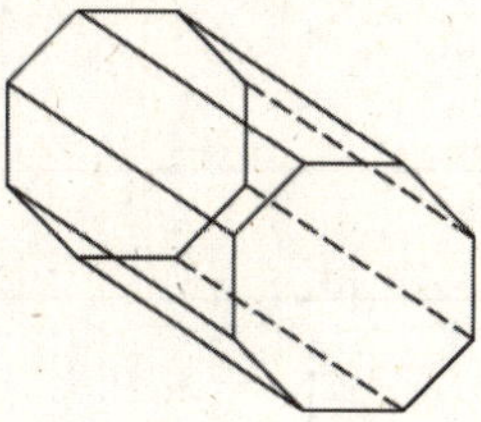

____

____

9. 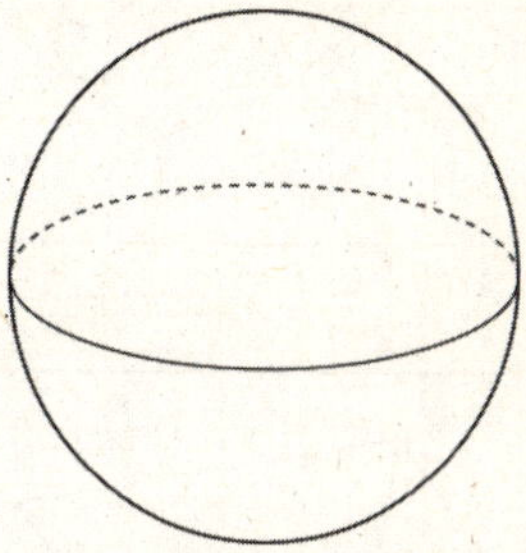

____

____

Name ______________________ Date ______________ Class ______________

LESSON 10-1

Practice B

Introduction to Three-Dimensional Figures

Identify the bases and faces of each figure. Then name the figure.

1. 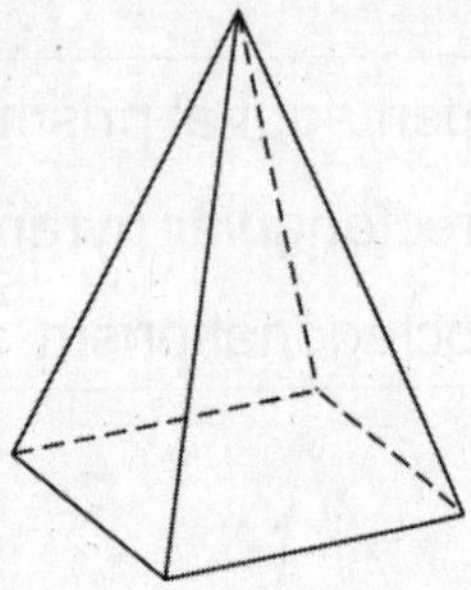

____

____

2. 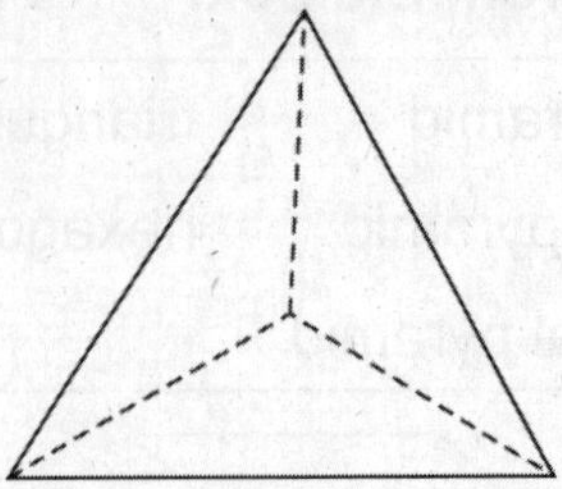

____

____

3. 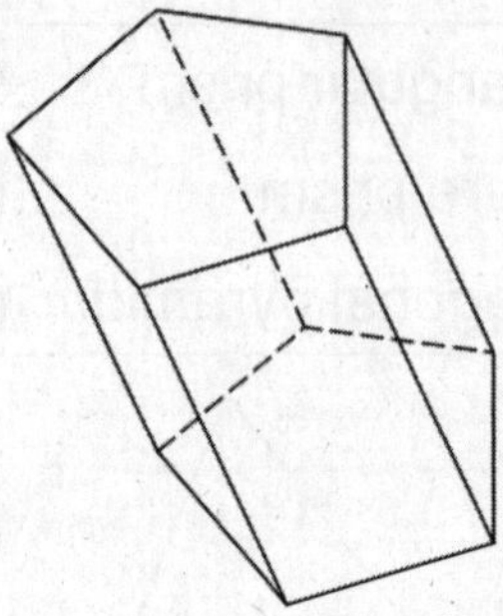

____

____

4. 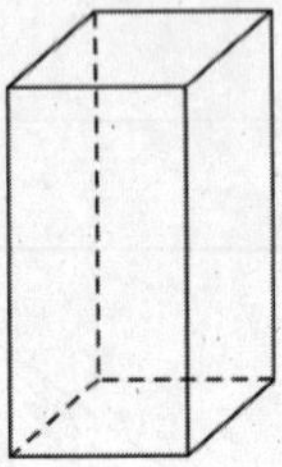

____

____

5. 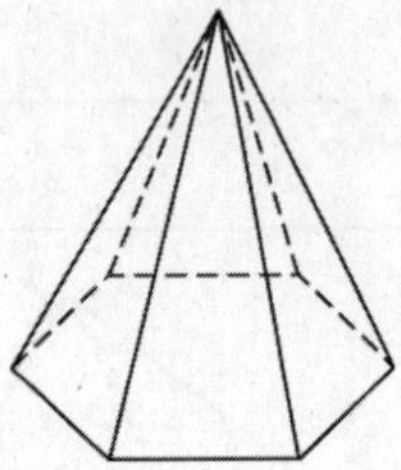

____

____

6. 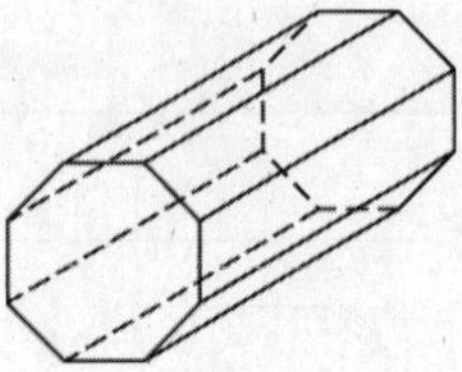

____

____

Classify each figure as a polyhedron or not a polyhedron. Then name the figure.

7. 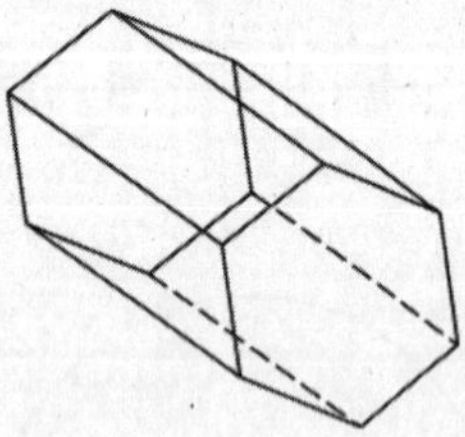

____

____

8. 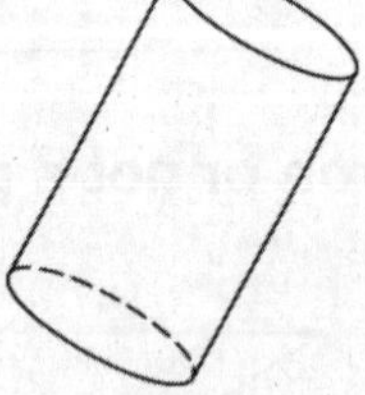

____

____

9. 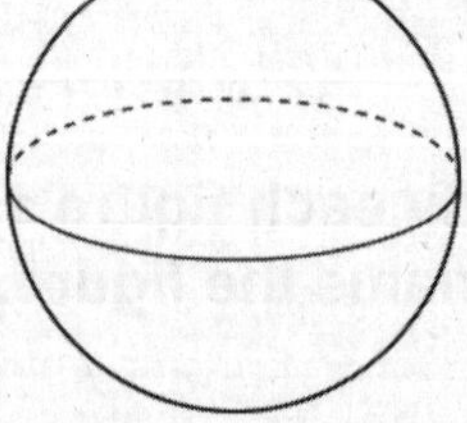

____

____

10. 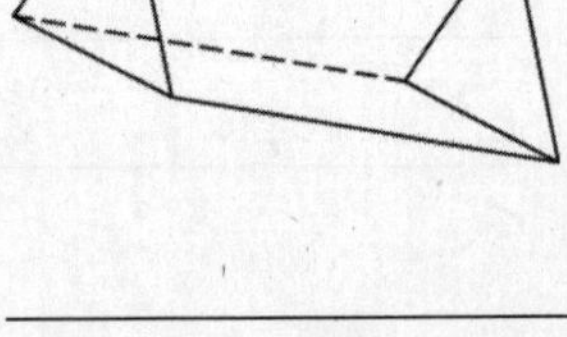

____

____

11. 

____

____

12. 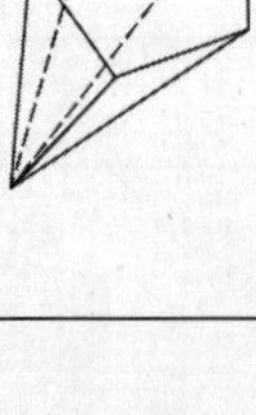

____

____

Name ______________________ Date ______________ Class ______________

LESSON 10-1

Practice C

Introduction to Three-Dimensional Figures

Identify the bases and faces of each figure. Then name the figure.

1. 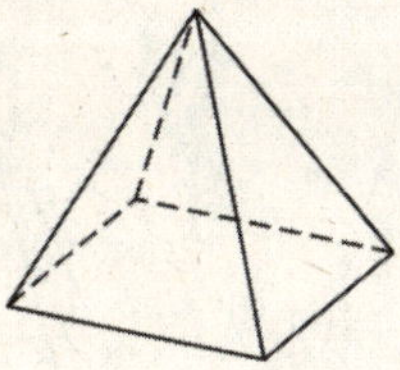

____

____

2. 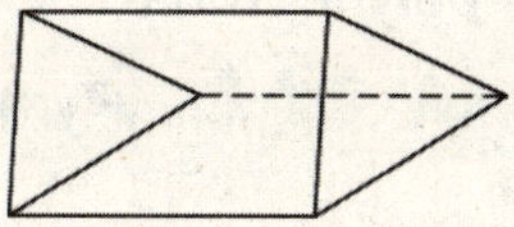

____

____

3. 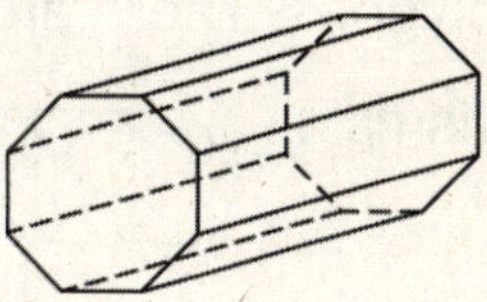

____

____

Classify each figure as a polyhedron or not a polyhedron. Then name the figure.

4. 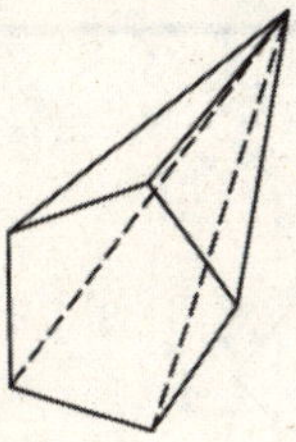

____

____

5. 

____

____

6. 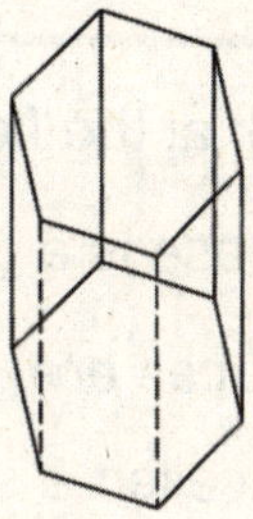

____

____

Identify the three-dimensional figure described.

7. three pairs of congruent rectangular sides ______________________

8. square base, four triangular faces ______________________

9. equilateral triangular base, three congruent equilateral triangular faces ______________________

10. two pentagonal bases, five parallelogram faces ______________________

11. two congruent circular bases ______________________

12. hexagonal base, six triangular faces ______________________

Name ______________________ Date ______________ Class ______________

LESSON 10-1

Review for Mastery

Introduction to Three-Dimensional Figures

A **polyhedron** is a three-dimensional figure whose faces are polygons. There are two types of polyhedrons.

Prisms

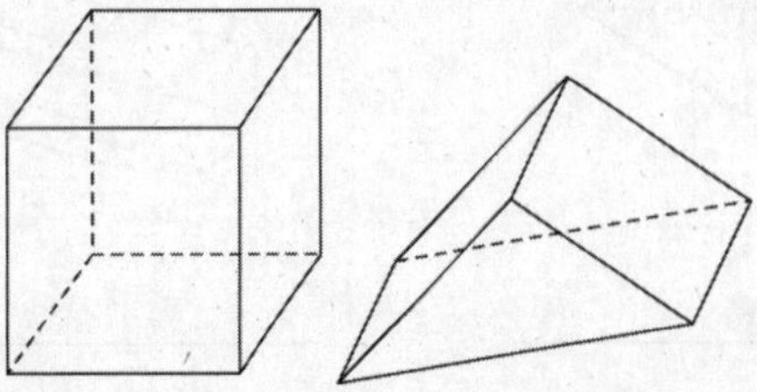

Pyramids

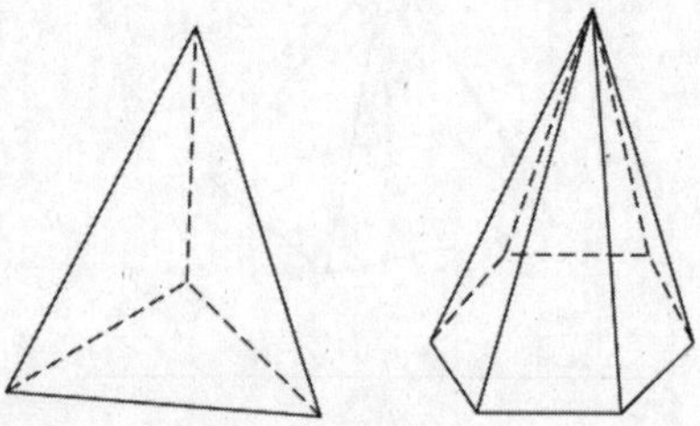

A **prism** has 2 bases that are congruent and parallel polygons. The other faces are parallelograms or rectangles.

A **pyramid** has one base that is a polygon. Its faces are triangles that meet at one vertex.

Prisms and pyramids are named by their bases.

1. Look at the figure at the right.

   Its base is a ______________.

   Its faces are ______________.

   It is called a square pyramid.

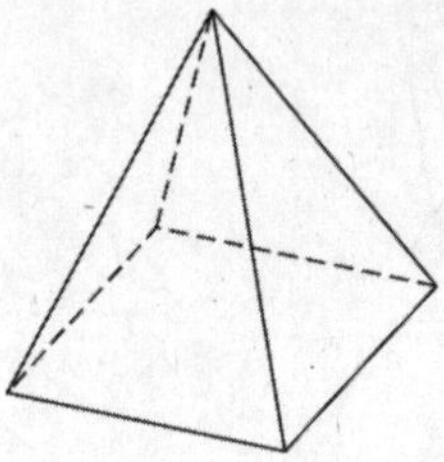

2. Look at the figure at the right.
   It has 2 congruent and parallel bases

   that are ______________.

   Its faces are ______________.

   It is called a pentagonal prism.

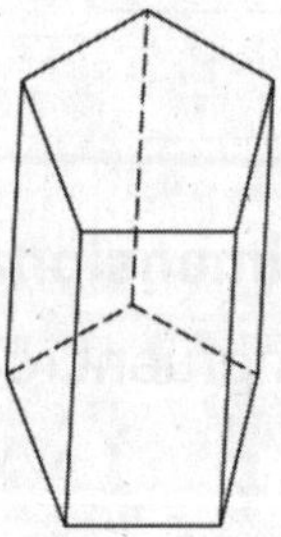

If a three-dimensional figure has a surface that is curved, it is not a polyhedron.

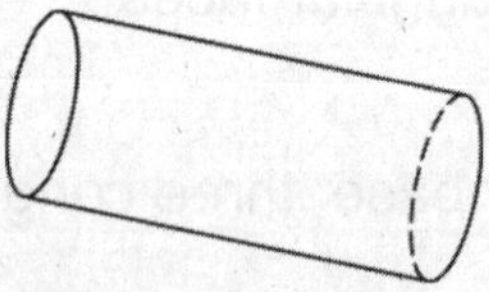

not a polyhedron

Is each figure a polyhedron?

3.

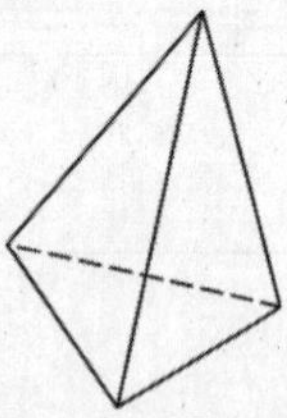

4.

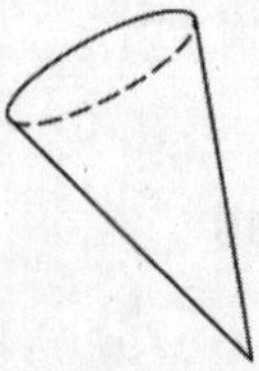

5.

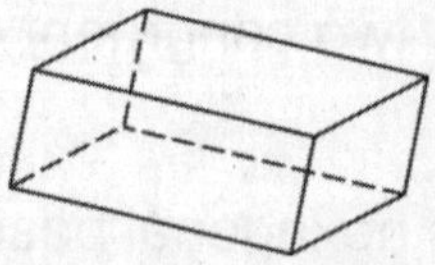

______________ ______________ ______________

Name ______________________ Date ______________ Class ______________

LESSON 10-1

Challenge

Platonic Solids

A three-dimensional figure in which all the faces are polygons is called a *polyhedron.* A polyhedron whose faces are all congruent regular polygons is called a *regular polyhedron*. Regular polyhedrons are called *Platonic solids*.

Below are patterns for four Platonic solids. Copy each pattern. You may enlarge the pattern if you like. Then cut out the pattern and fold it to build a model of the solid. Use your models to answer the questions.

Hexahedron

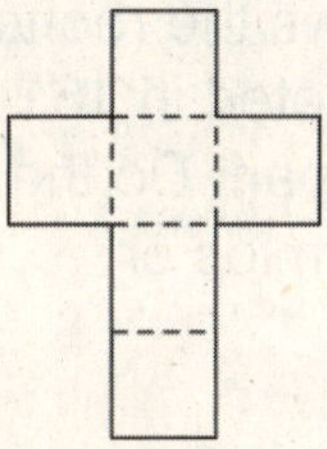

Tetrahedron

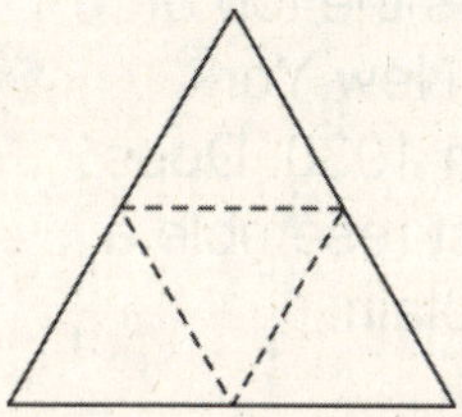

Octahedron

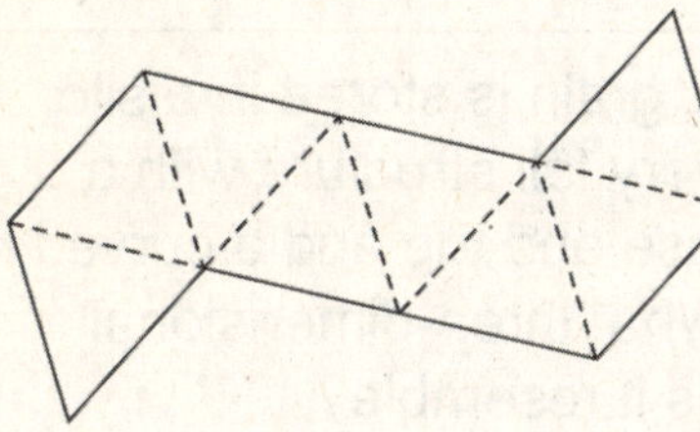

Dodecahedron

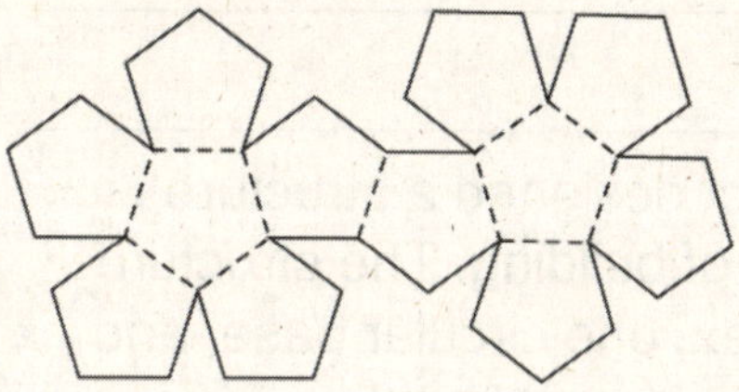

1. How many faces does a hexahedron have? What is another name for a hexahedron?

____

2. How many faces does a tetrahedron have? What is another name for a tetrahedron?

____

3. How many faces does an octahedron have?

____

4. How many faces does a dodecahedron have?

____

Name ______________________ Date ____________ Class ____________

LESSON 10-1

Problem Solving

Introduction to Three-Dimensional Figures

Write the correct answer.

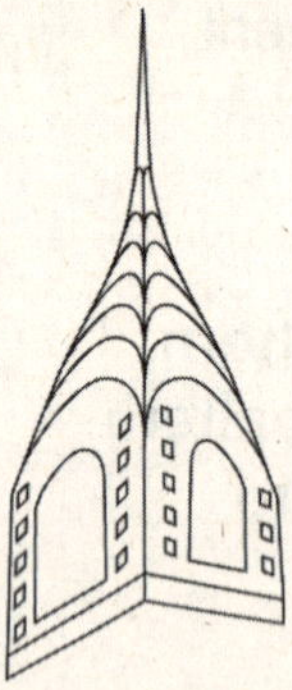

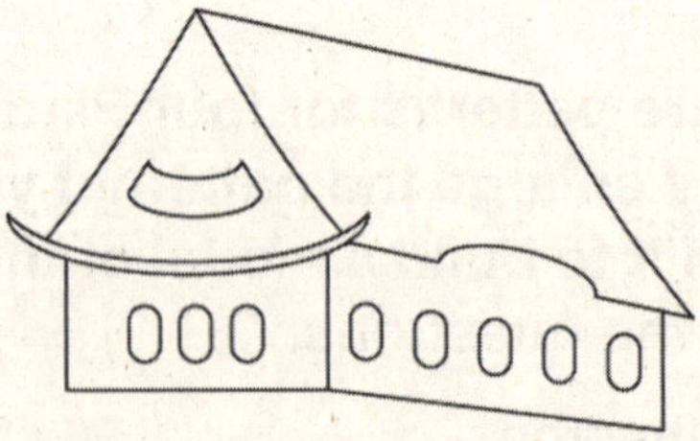

1. The picture above shows the top of the Chrysler Building in New York City. It was completed in 1930. Does the top of the tower most resemble a prism or a pyramid? Explain.

____

____

____

2. The picture above shows the rooftop of Himeji Castle, completed in 1614 in Donjon, Himeji City, Japan. Do the rooftops resemble pyramids or prisms? Explain.

____

____

____

3. An architect designed a structure for the top of building. The structure has a vertex, one circular base, and a curved surface. What three-dimensional figure is it?

____

4. On a farm, grain is stored in a silo. This is a very tall structure with a circular base and top and a curved surface. What three-dimensional figure does it resemble?

____

Choose the letter of the correct answer.

5. James put two blocks together to build the figure shown. Identify the two figures he used.

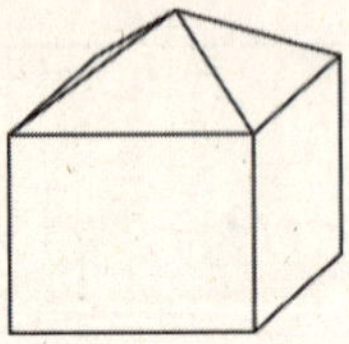

A two pyramids

B a pyramid and a prism

C a pyramid and a cone

D two prisms

6. Jaime constructs a figure that has one rectangular base and four triangular faces. What is the figure?

F a cone

G a triangular pyramid

H a rectangular pyramid

J a rectangular prism

7. The shape of a log is most like which figure?

A cylinder

B pyramid

C prism

D cone

Name ______________________ Date ______________ Class ______________

LESSON 10-1

Reading Strategies

Focus on Vocabulary

A **polyhedron** is a three-dimensional figure with flat sides called **faces.** In a polyhedron, all of the faces are polygons.

POLYHEDRONS	
Prism 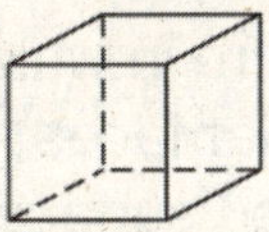 • Three-dimensional figure • Parallel, congruent faces • Two opposite faces are called "bases."	**Pyramid** 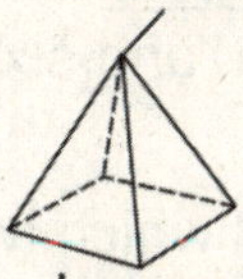 • Three-dimensional figure • Triangular faces meet at a common vertex. • Face opposite vertex is the base.

1. What name is given to a three-dimensional figure with flat sides? ____________

2. What do you call a flat side of a three-dimensional figure? ____________

3. Which type of polyhedron has two parallel, congruent faces? ____________

4. Which type of polyhedron has a vertex and a face at opposite ends? ____________

5. What do you call the two parallel, congruent faces of a prism? ____________

6. What do you call the face opposite the vertex in a pyramid? ____________

7. Why isn't a cone a polyhedron? ____________

____

8. Name another three-dimensional figure that is not a polyhedron.

____

Name ____________________ Date ____________ Class ____________

LESSON 10-1

Puzzles, Twisters & Teasers

You Are Now Entering the X-Word Dimension

Solve the crossed puzzle.

Across

3. A _______ is a three-dimensional figure whose faces are all polygons.
6. A hexagonal _______ has one base that is a hexagon, and six faces that are triangles.
7. a line segment at which two faces of a prism meet
8. a polyhedron with two parallel congruent bases and all other faces being parallelograms
10. The face opposite the _______ is the base of the pyramid.

Down

1. A _______ has one circular face and a curved surface that comes to a point called the vertex.
2. a figure in which two congruent circular faces are connected by a curved surface
4. A _______ prism has six faces that are rectangles.
5. A prism is named for its two congruent parallel _______.
9. Each surface of a polyhedron is called a _______.

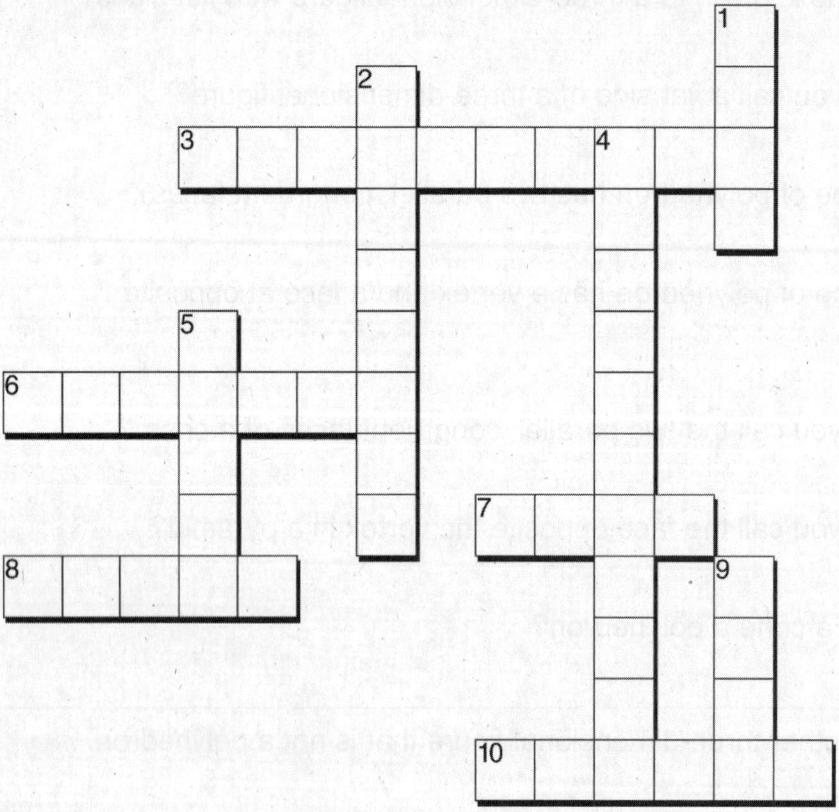

Name ______________________ Date ______________ Class ______________

LESSON 10-2

Practice A

Volume of Prisms and Cylinders

Find the volume of each figure. Choose the letter for the best answer.

1.

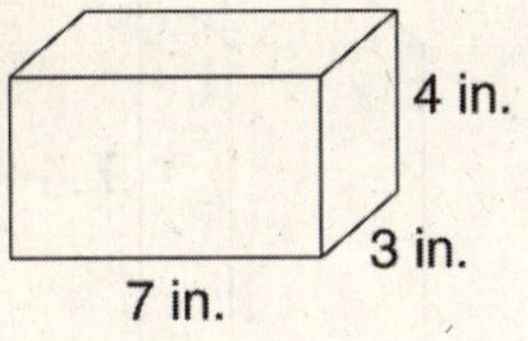

A 14 in^3 C 28 in^3

B 21 in^3 D 84 in^3

2.

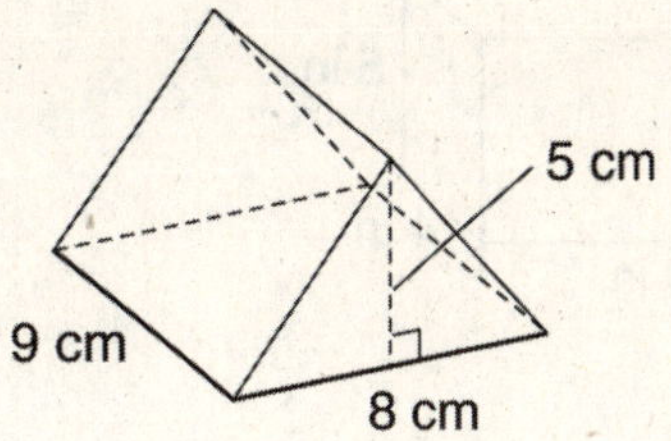

F 40 cm^3 H 180 cm^3

G 72 cm^3 J 360 cm^3

3.

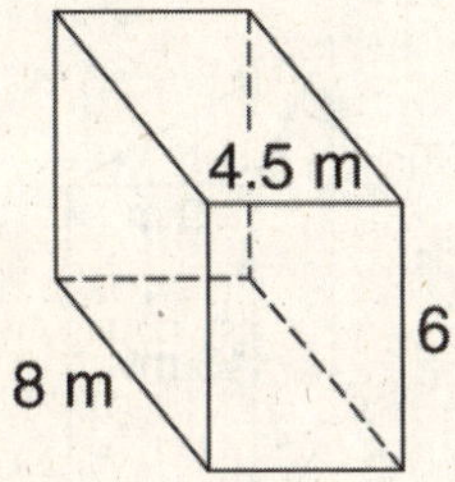

A 48 m^3 C 216 m^3

B 162 m^3 D Not Here

4.

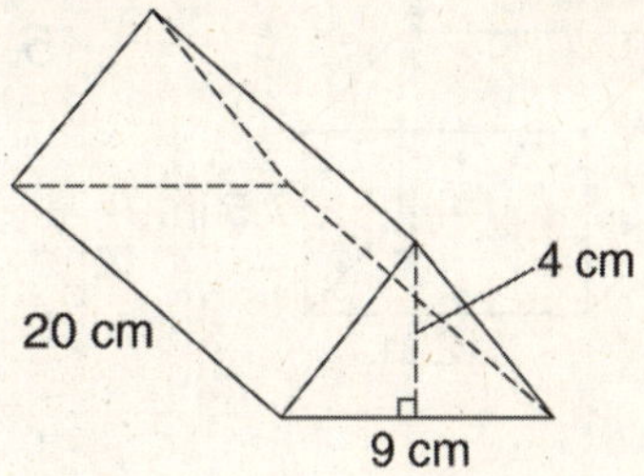

F 60 cm^3 H 720 cm^3

G 360 cm^3 J 810 cm^3

5. A juice can is shaped like a cylinder. It is 12 centimeters wide and 16 centimeters tall. Find its volume to the nearest whole number. Use 3.14 for π.

____

Find the volume of the composite figure to the nearest tenth. Use 3.14 for π.

6.

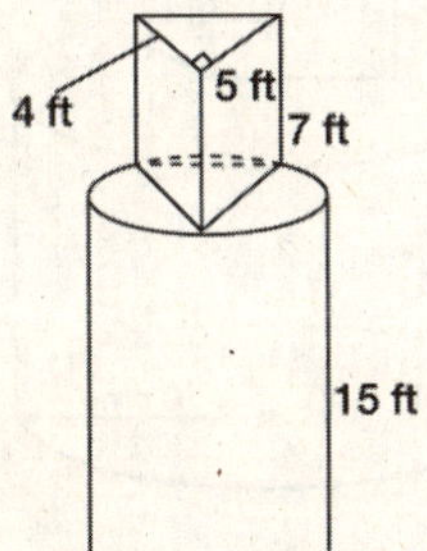

7.

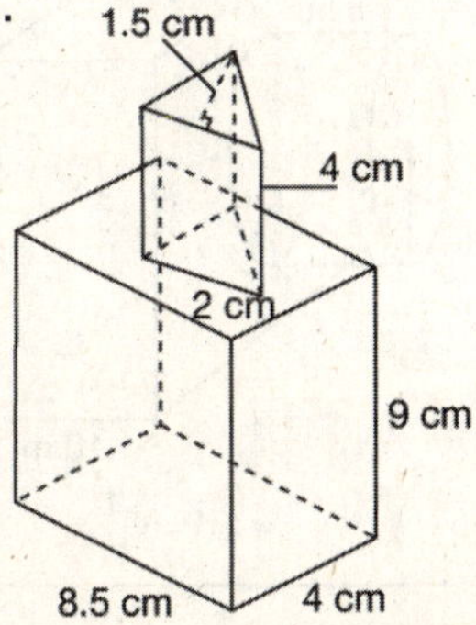

8.

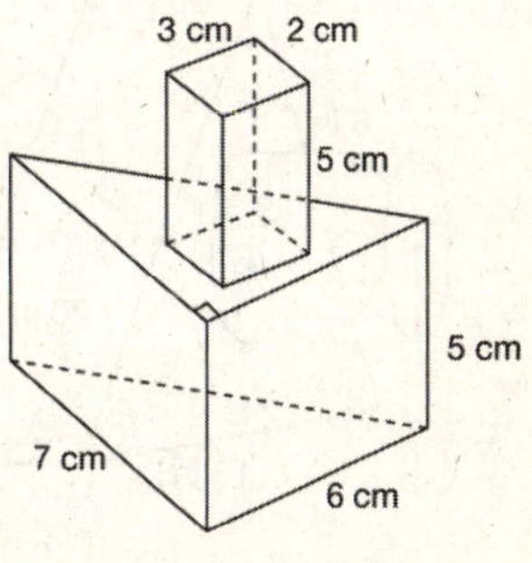

______________________ ______________________ ______________________

Name ______________________ Date ______________ Class ______________

LESSON 10-2

Practice B

Volume of Prisms and Cylinders

Find the volume of each figure.

1.

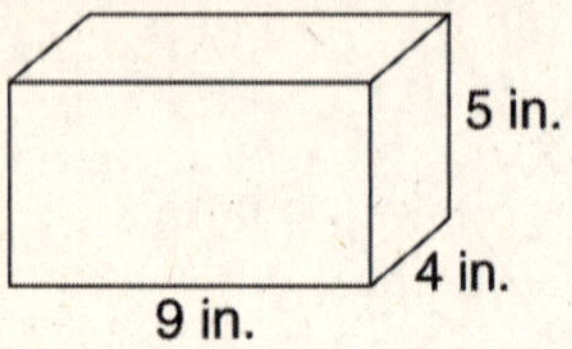

2.

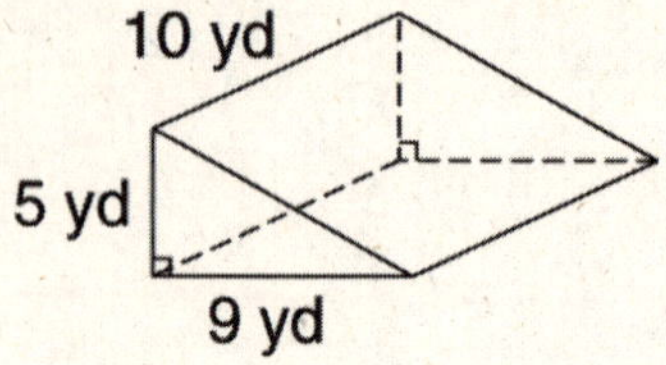

3.

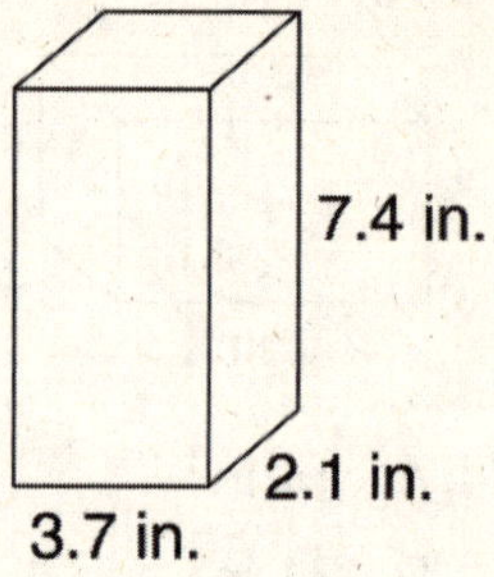

______________ ______________ ______________

4.

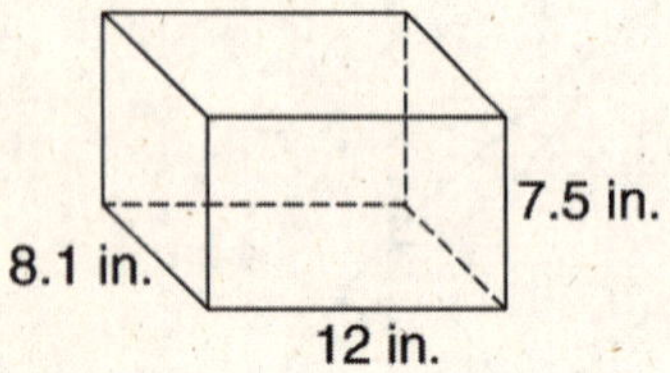

5.

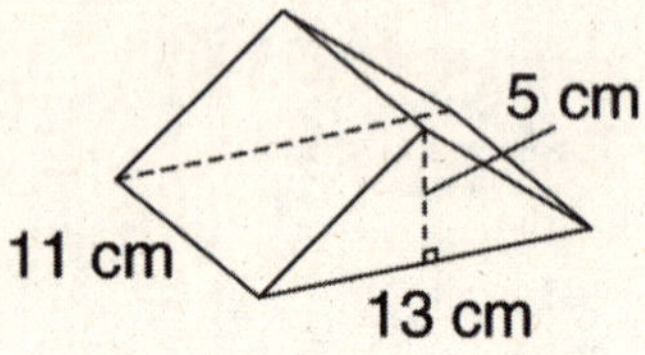

6.

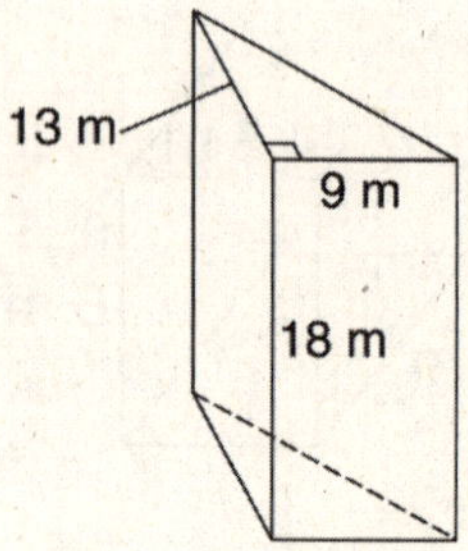

______________ ______________ ______________

7. A travel mug is shaped like a cylinder. It is 9 centimeters wide and 15 centimeters tall. Find its volume to the nearest tenth. Use 3.14 for π.

____

Find the volume of the composite figure to the nearest tenth. Use 3.14 for π.

8.

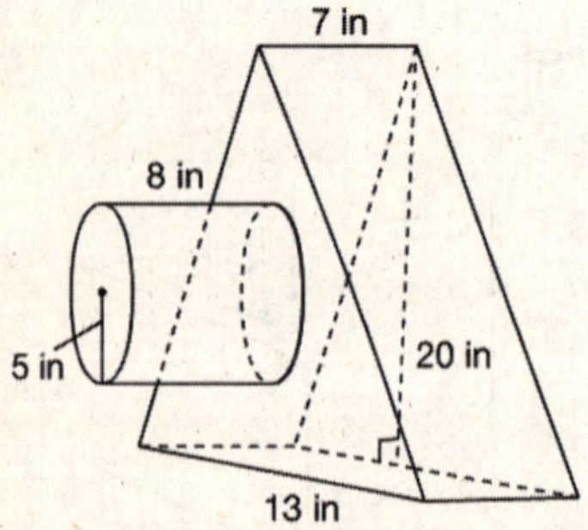

9.

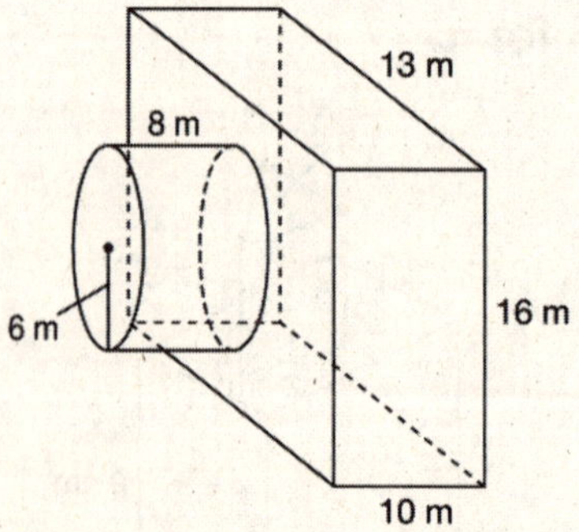

10.

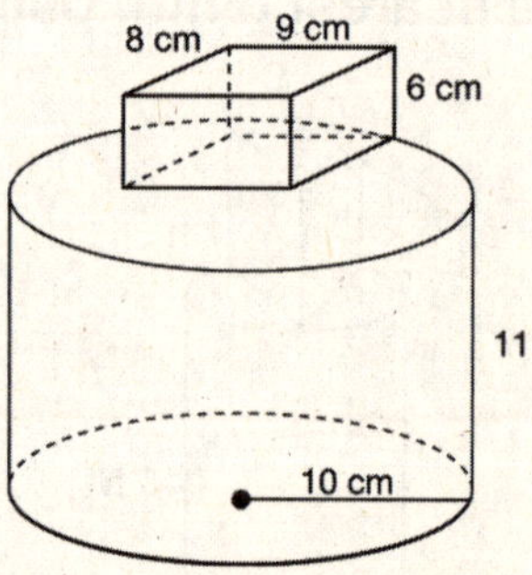

______________ ______________ ______________

Name ______________________ Date ____________ Class ____________

LESSON 10-2

Practice C

Volume of Prisms and Cylinders

Find the volume of each figure to the nearest tenth.

1.
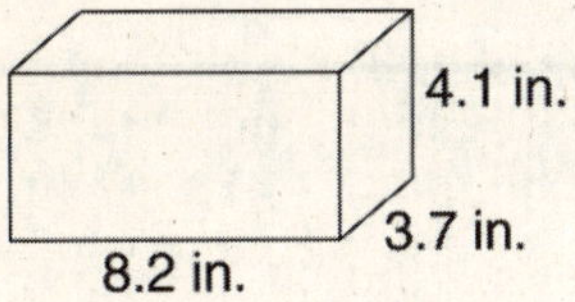

2.
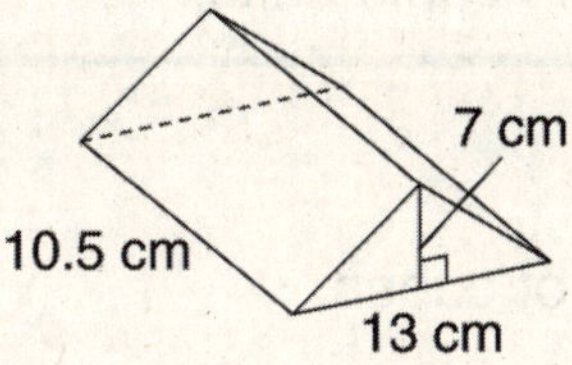

3.
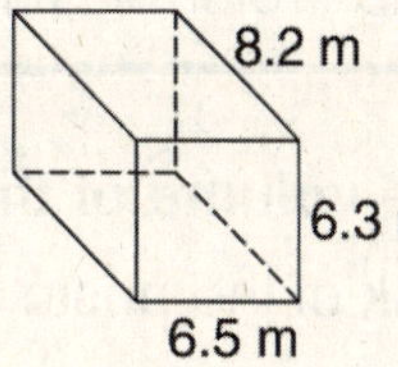

4.
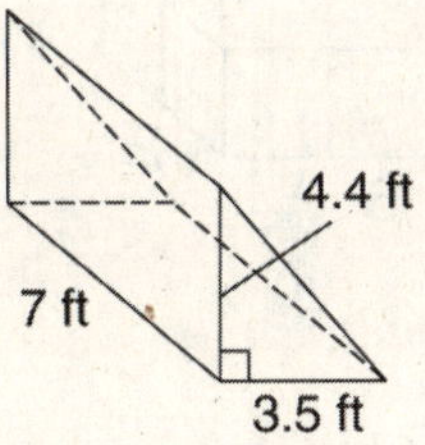

5.
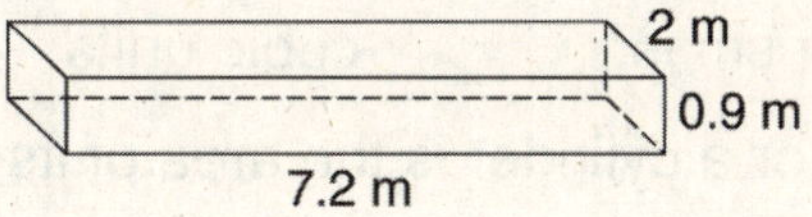

6.
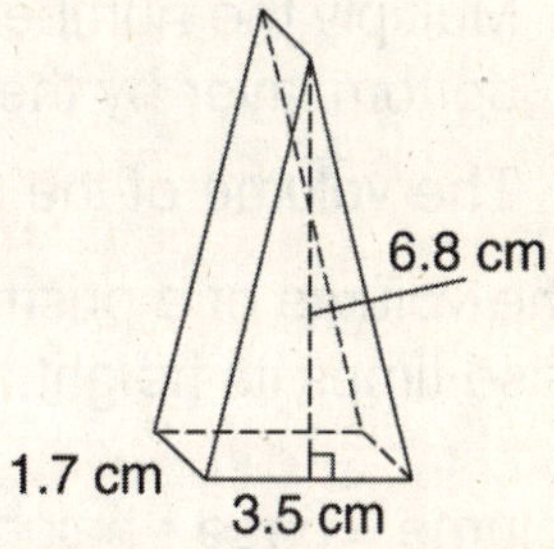

7. A vase is in the shape of a cylinder with a diameter of 5 inches and a height of 9 inches. What is the volume of the vase? Use 3.14 for π.

8. A glass is shaped like a cylinder. The glass has a volume of 602.88 centimeters and a radius of 4 centimeters. What is the height?

9. The base of a triangular prism is a right triangle with hypotenuse 13 inches long and one leg 12 inches long. The height of the prism is 8 inches. What is the volume of the prism?

Find the volume of the composite figure to the nearest tenth. Use 3.14 for π.

10.
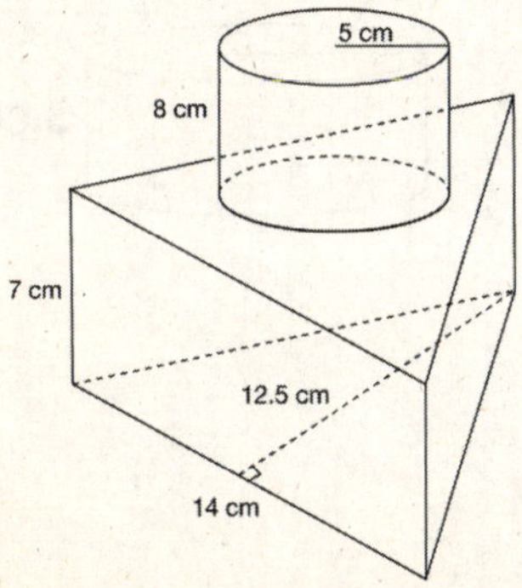

11.
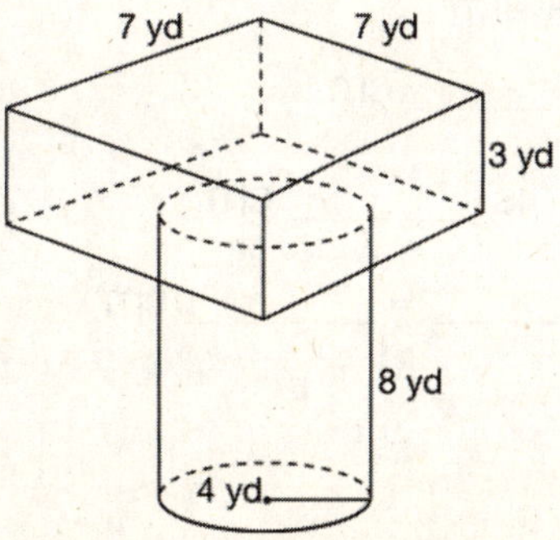

12.
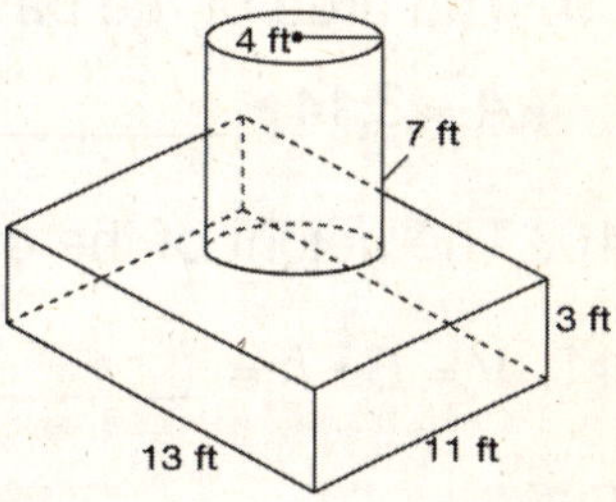

Name ______________________ Date ______________ Class ______________

LESSON 10-2

Review for Mastery

Volume of Prisms and Cylinders

The **volume** of a three-dimensional figure is the amount of space it takes up. Volume is measured in cubic units.

Find the volume of the prism.

1. Think of the prism as layers of cubes.

   There are _______ cubes in the bottom layer.

2. There are _______ layers of cubes.

3. Multiply the number of cubes in the bottom layer by the number of layers.

   The volume of the prism is _______ cubic units.

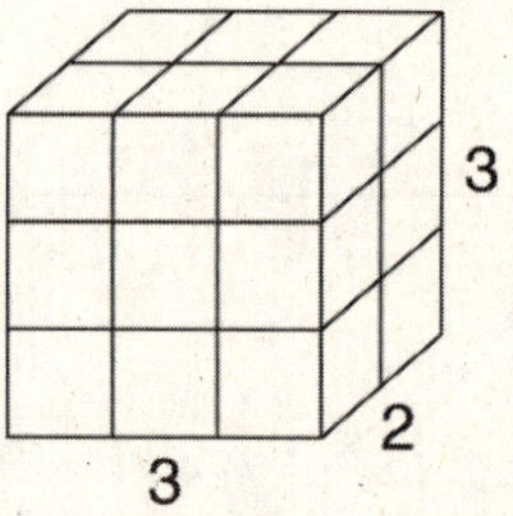

The volume of a prism or a cylinder is the area of its base times its height.

volume = base • height, or $V = B \cdot h$

Find the volume of the prism.

4. What is the shape of the base? ____________

5. The area of the base is $B = \frac{1}{2}bh$

   $B = \frac{1}{2} \cdot$ _______ • _______ = _______ in^2

6. The height of the prism is _______ in.

7. $V = B \cdot h =$ _______ • _______ = _______ in^3

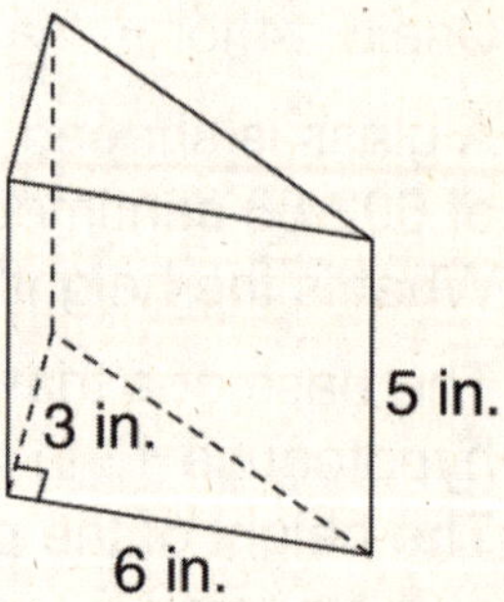

Find the volume of the cylinder to the nearest whole number.

8. What is the shape of the base? ____________

9. The area of the base is $A = \pi r^2$.

   $A = 3.14 \cdot$ _______ ^{2} = _______ cm^2

10. The height of the cylinder is _______ cm.

11. $V = B \cdot h =$ _______ • _______ = _______ cm^3

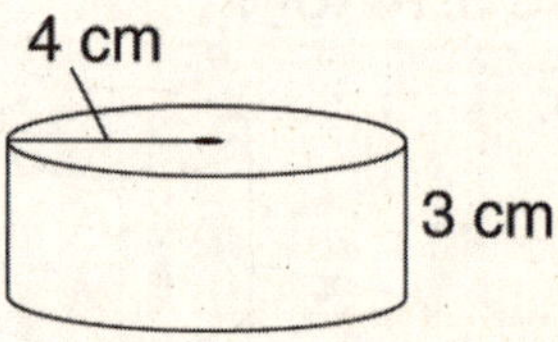

Name ______________________ Date ______________ Class ______________

LESSON 10-2

Challenge

Painted Faces

A cube has six sides, or faces.
Each of the six faces is a square.

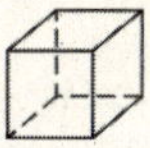

This cube measures 2 units on each edge. The faces of the cube are painted.

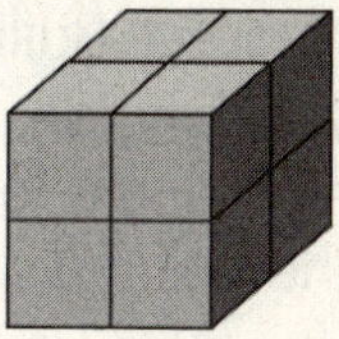

1. How many small cubes make up the large cube? ______________
2. How many small cubes are painted on 3 faces? on 2 faces? on 1 face? not at all?

____

This cube measures 3 units on each edge.

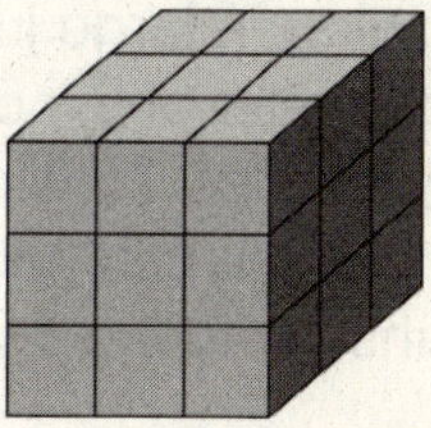

3. How many small cubes make up the large cube? ______________
4. If the large cube is painted, how many small cubes will be painted on 3 faces? on 2 faces? on 1 face? not at all?

____

This cube measures 4 units on each edge.

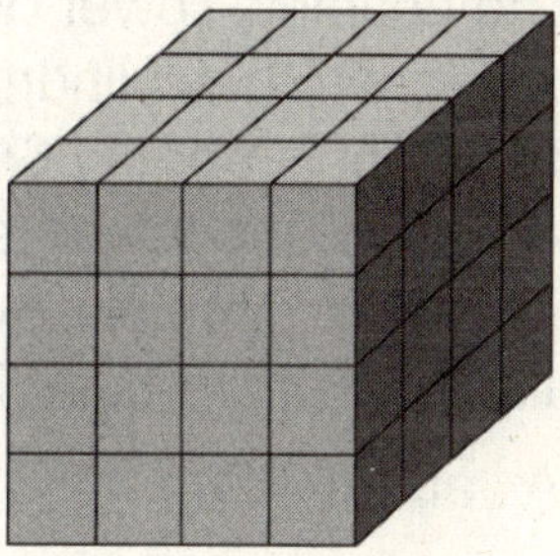

5. How many small cubes make up the large cube? ______________
6. If the large cube is painted, how many small cubes will be painted on 3 faces? on 2 faces? on 1 face? not at all?

____

7. Complete the table below by entering the number of small cubes that will be painted on the given number of faces.

Size of Large Cube	Painted on 3 Faces	Painted on 2 Faces	Painted on 1 Face	Painted on 0 Faces
2 × 2 × 2				
3 × 3 × 3				
4 × 4 × 4				
5 × 5 × 5				

8. Look at your table in Exercise 7. What is the rule for the number of cubes painted on 3 faces? ______________

Name ______________________ Date ______________ Class ______________

LESSON 10-2

Problem Solving

Volume of Prisms and Cylinders

Write the correct answer.

1. The eight Corinthian columns at the National Building Museum in Washington, DC, are each 75 feet high and 8 feet in diameter. What is the volume of each column?

   ____

2. A cubic centimeter holds 1 milliliter of liquid. How many liters of water to the nearest tenth are required to fill a fish tank that is 24 centimeters high, 28 centimeters long, and 36 centimeters wide?

   ____

3. There are 231 cubic inches in a gallon. A large juice can has a diameter of 6 inches and a height of 10 inches. How many gallons of juice does the can hold? Round your answer to the nearest tenth.

   ____

4. A small gift box that holds a ring is shaped like a cube. The box measures 1.4 inches on each side. What is the volume of the gift box? Round your answer to the nearest tenth.

   ____

Choose the letter of the correct answer.

5. The Leaning Tower of Pisa in Italy appears to be cylindrical in shape. Its height is about 56 meters. If the volume of the tower is about 9,891 cubic meters, what is the diameter of the base?

   A about 3.5 m　　C about 15 m
   B about 7 m　　D about 20 m

6. A bricklayer is building a brick rectangular post to anchor a mailbox. The post is 3 feet tall, 2 feet deep, and 2 feet wide. Each brick is 3 inches by 6 inches by 3 inches. How many bricks does he need?

   F 12 bricks　　H 197 bricks
   G 54 bricks　　J 384 bricks

7. The average stone on the lowest level of the Great Pyramid in Egypt was a rectangular prism 5 feet long by 5 feet high by 6 feet deep and weighed 15 tons. What was the volume of the average stone?

   A 1,800 ft^3　　C 150 ft^3
   B 1,800 ft^2　　D 150 ft^2

8. A cylindrical barrel is 2.8 feet in diameter and 8 feet high. If a cubic foot holds about 7.5 gallons of liquid, how many gallons of water will this barrel hold?

   F about 1,477 gal
   G about 369 gal
   H about 150 gal
   J about 470 gal

Name ______________________ Date ______________ Class ______________

LESSON 10-2

Reading Strategies

Use a Visual Aid

Think of the **volume** of a solid figure as the number of cubic units inside the figure.

one cubic unit

Each cube represents one cubic unit.

1. What is the height of the prism in the figure? ____________

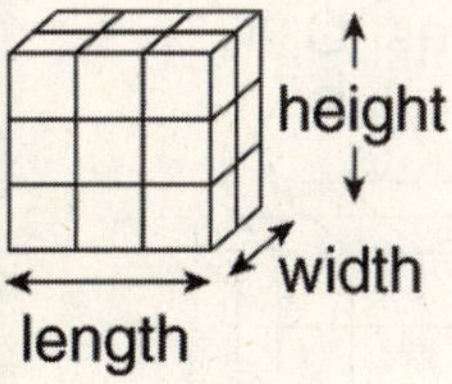

2. What is the width of the prism? ____________
3. What is the length of this prism? ____________

Multiply the length, width, and height of a prism to find its volume in cubic units.

Volume = length • width • height

↓ ↓ ↓ ↓

$V = \ell \cdot w \cdot h$

4. Multiply the length, width, and height of the prism above to find the volume. What is the volume? ____________

Use this rectangular prism to complete Exercises 5–7.

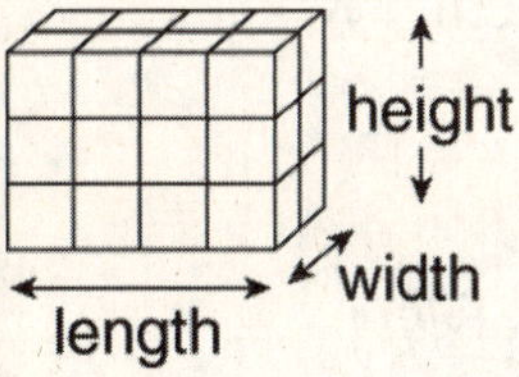

5. How long is the prism? How wide is the prism? ____________
6. What is the height of the prism? ____________
7. Use the formula $V = \ell \cdot w \cdot h$ to find the volume of this prism.

____

Name ______________________ Date ______________ Class ______________

LESSON **10-2**

Puzzles, Twisters & Teasers

Crack the Code!

Find how many cubes each prism holds. Circle the letter of the correct answer.

1. 16 cubes P
   12 cubes L

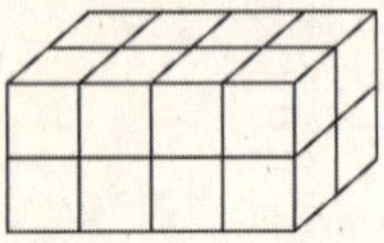

2. 12 cubes Y
   16 cubes P

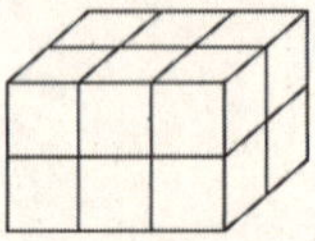

3. 12 cubes Y
   24 cubes U

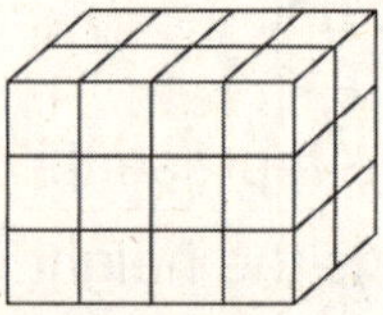

4. 54 cubes A
   45 cubes O

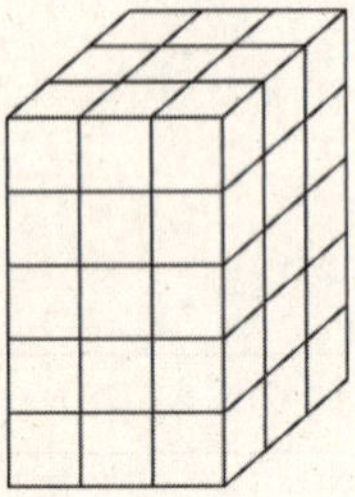

5. 80 cubes E
   60 cubes I

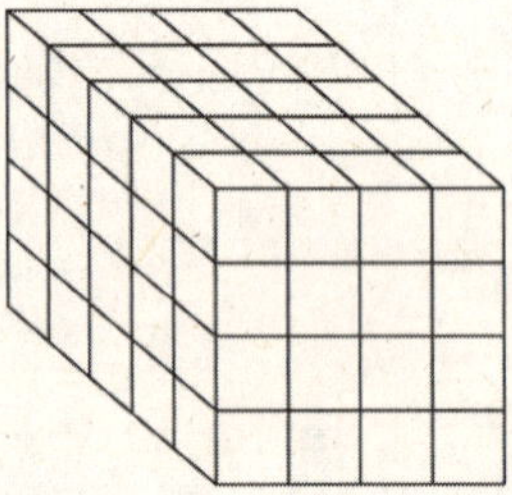

6. 20 cubes C
   16 cubes S

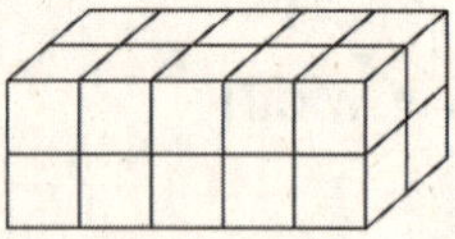

Find the volume of each figure to the nearest whole number. Circle the letter of the correct answer.

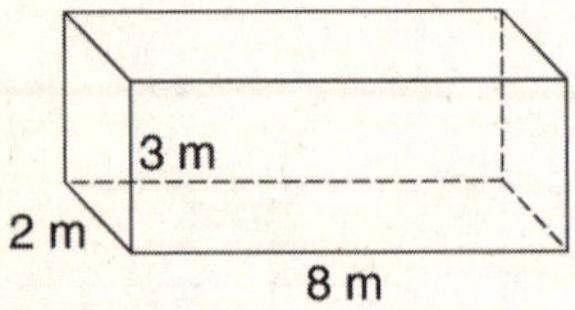

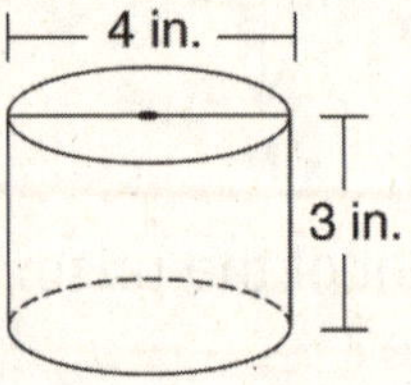

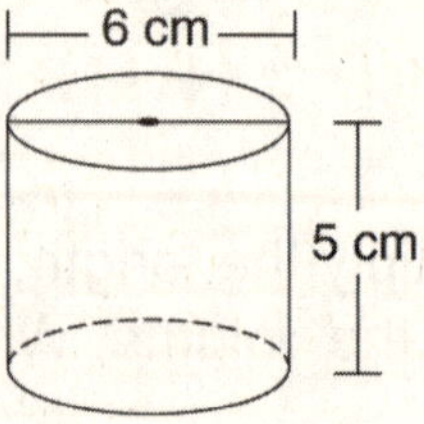

7. 46 m^3 N
   48 m^3 M

8. 38 in^3 R
   151 in^3 H

9. 47 cm^3 P
   141 cm^3 K

Write the letters above the answers to solve the riddle.

What did the egg say to the clown?

___ ___ ___   ___ ___ A ___ ___
12 45 24   20 38   20 141

___ ___   ___ ___.
48 80   24 16

Name ______________________ Date ______________ Class ______________

LESSON 10-3

Practice A

Volume of Pyramids and Cones

1. Write the formula for the volume of a pyramid. ______________

2. Write the formula for the volume of a cone. ______________

Find the volume of each pyramid or cone to the nearest whole number. Use 3.14 for π. Cross out each number in the box that matches a volume.

2,826	34	8	500	360	36
471	6	33	7	38	540

3.
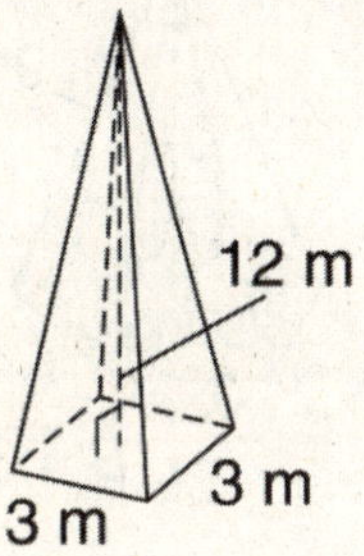

4.
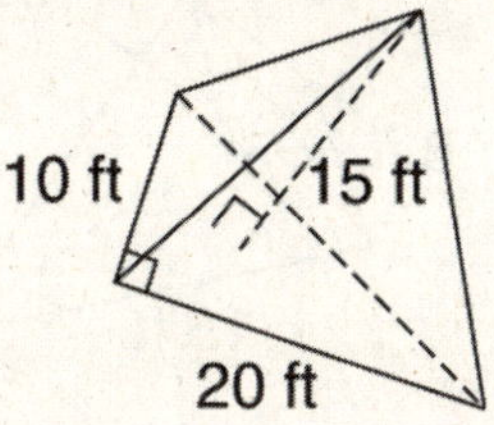

5.
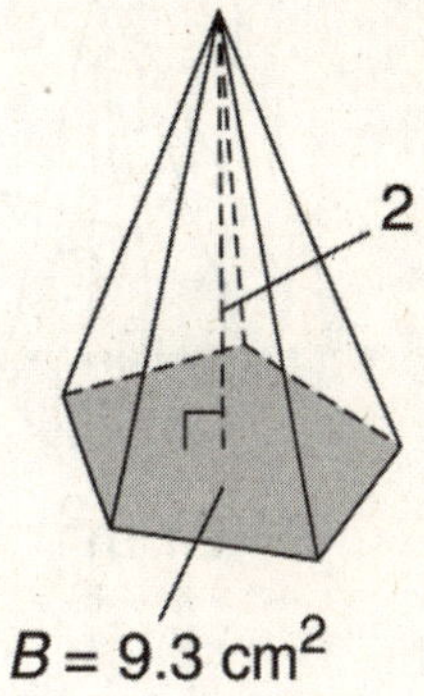

6.
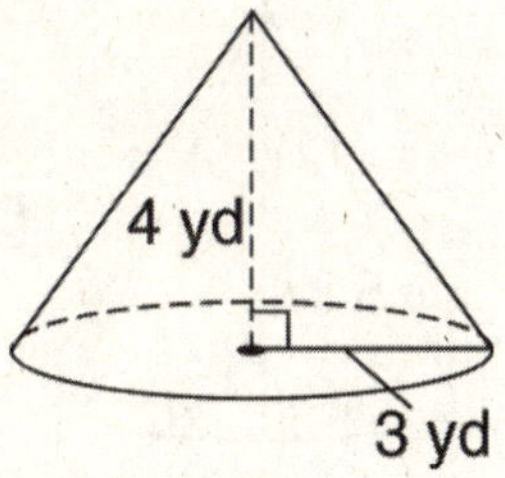

7.
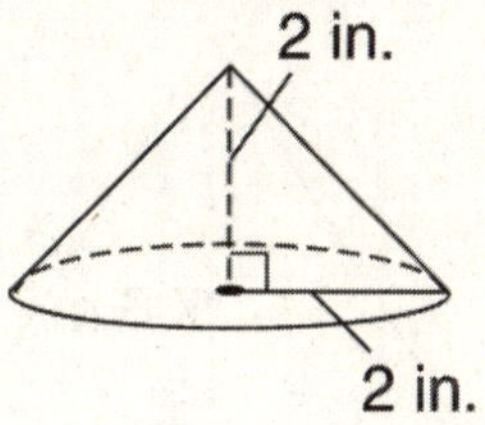

8.
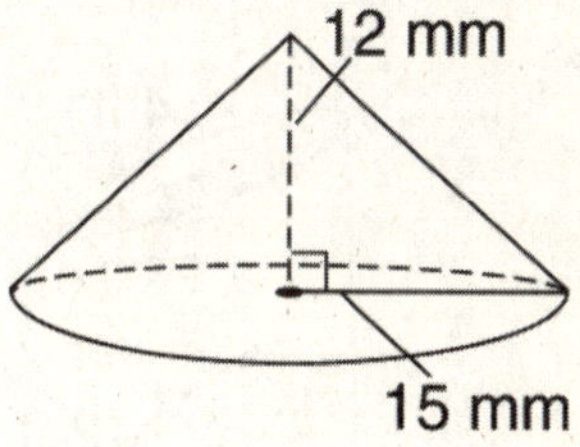

9.
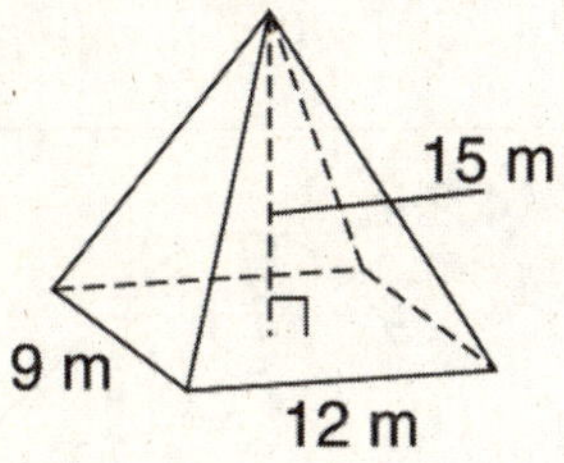

10.
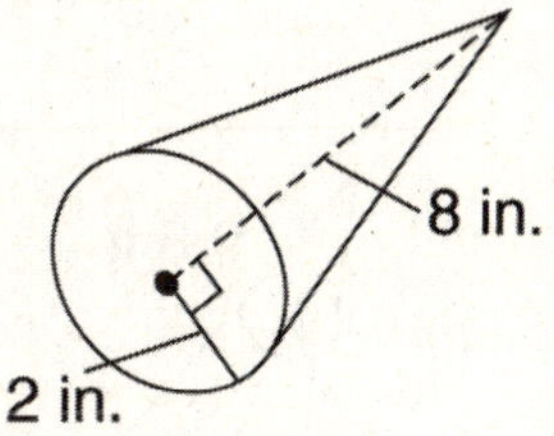

11.
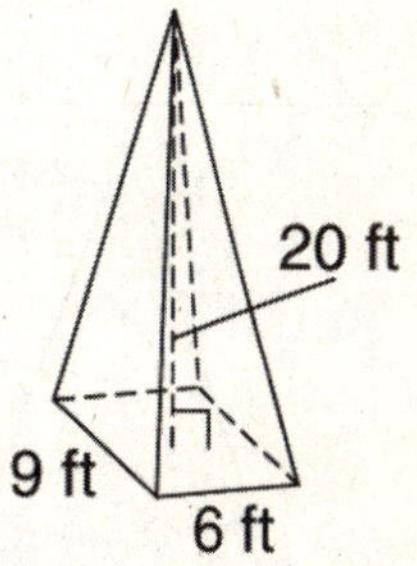

Name ______________________ Date ______________ Class ______________

LESSON 10-3

Practice B

Volume of Pyramids and Cones

Find the volume of each pyramid to the nearest tenth.

1. 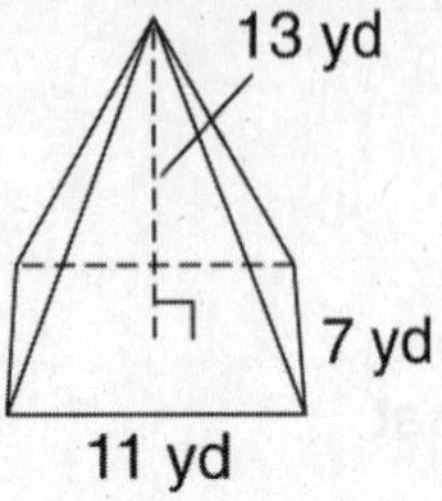

2. 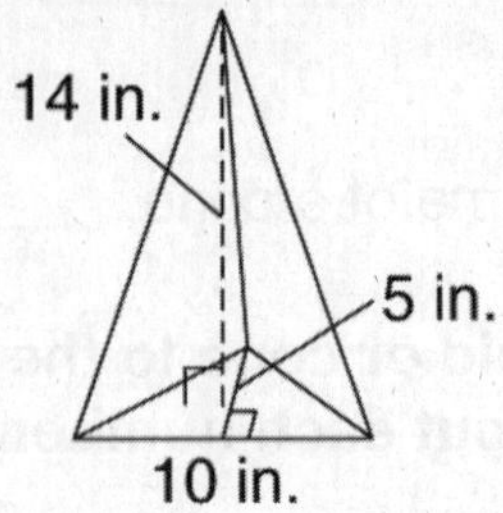

3. 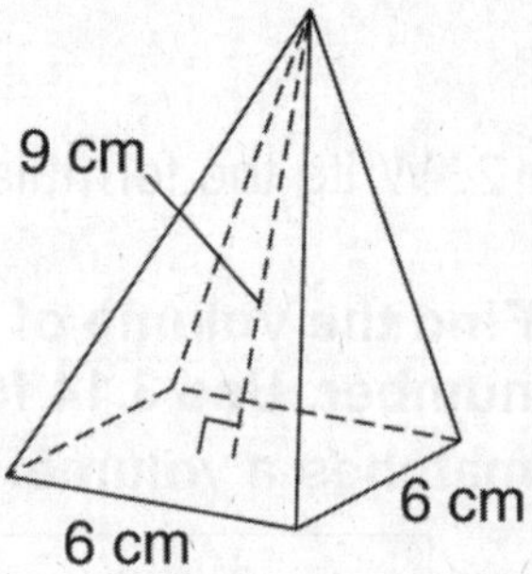

______________ ______________ ______________

4. 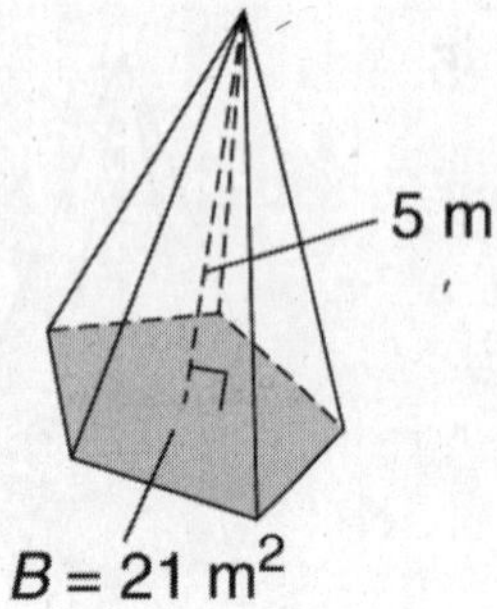

5. 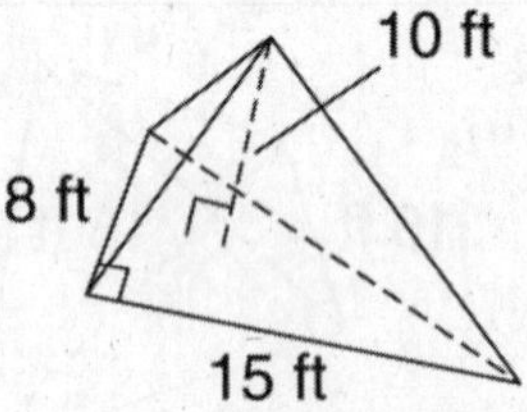

6. 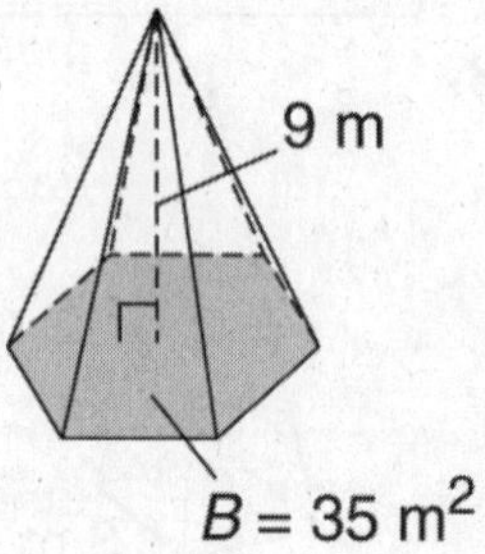

______________ ______________ ______________

Find the volume of each cone to the nearest tenth. Use 3.14 for π.

7. 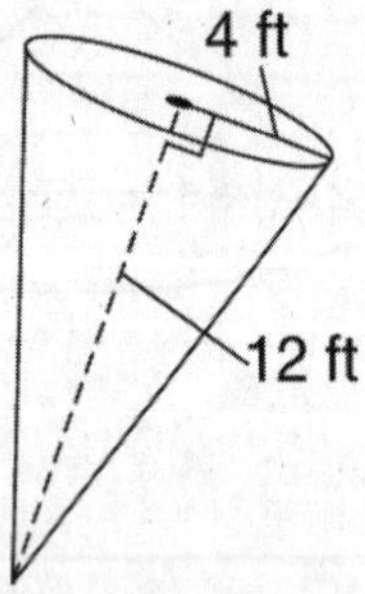

8. 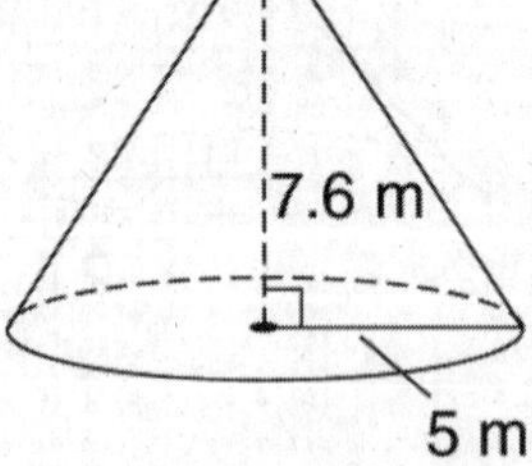

9. 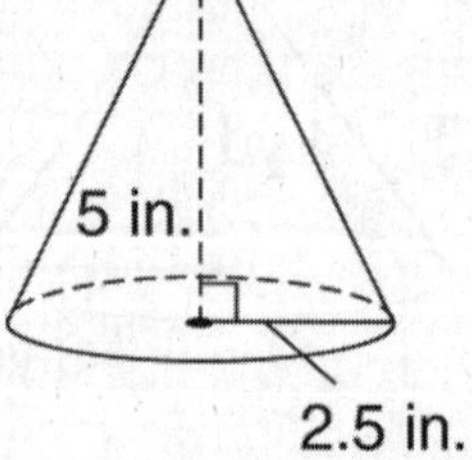

______________ ______________ ______________

10. 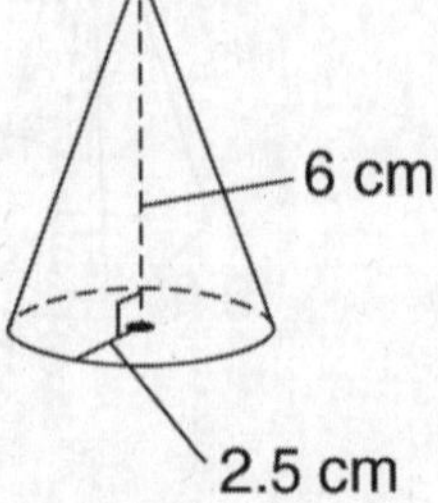

11. 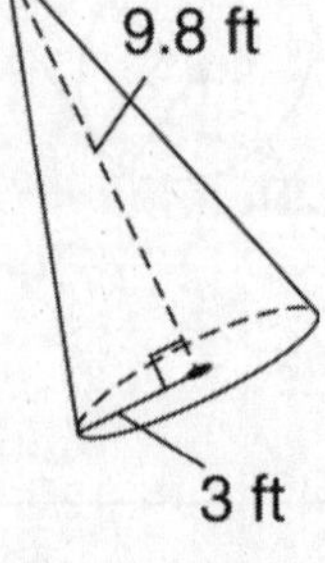

12. 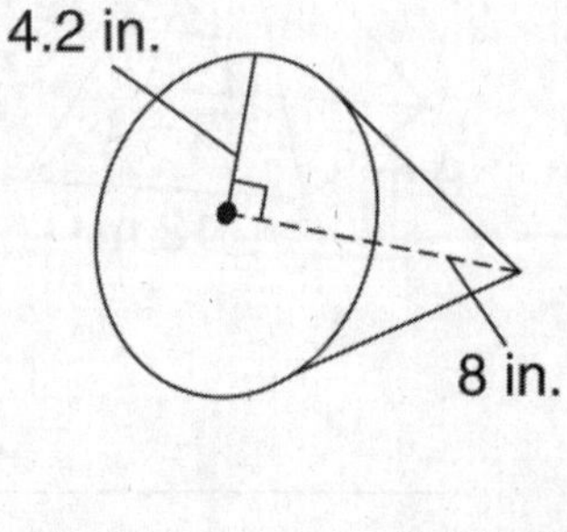

______________ ______________ ______________

Name ________________ Date ________________ Class ________________

LESSON 10-3

Practice C

Volume of Pyramids and Cones

Find the volume of each pyramid to the nearest tenth.

1.

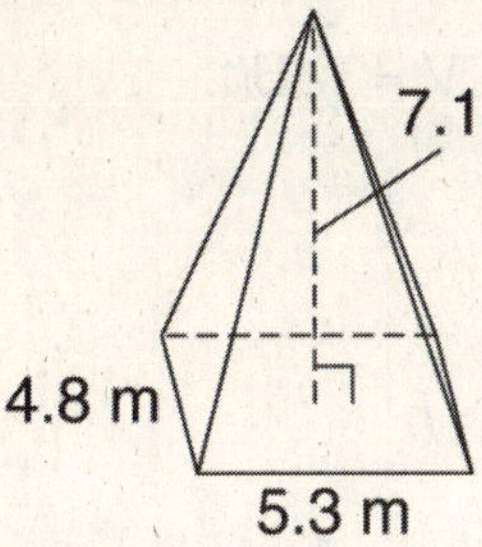

____

2.

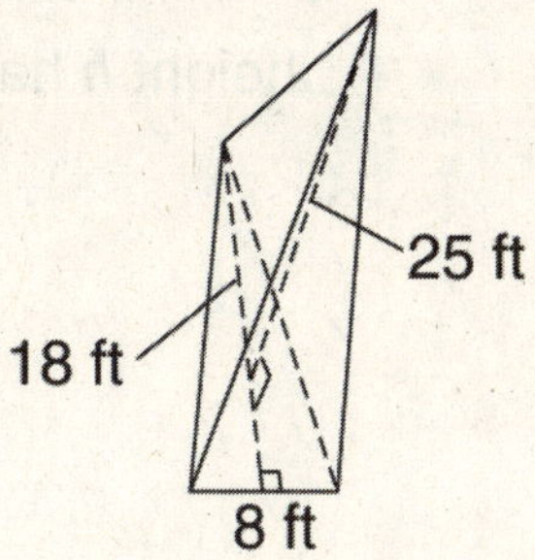

____

3.

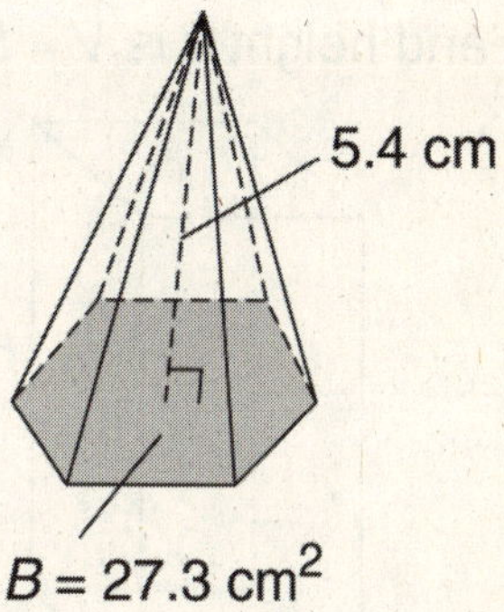

____

Find the volume of each cone to the nearest tenth. Use 3.14 for π.

4.

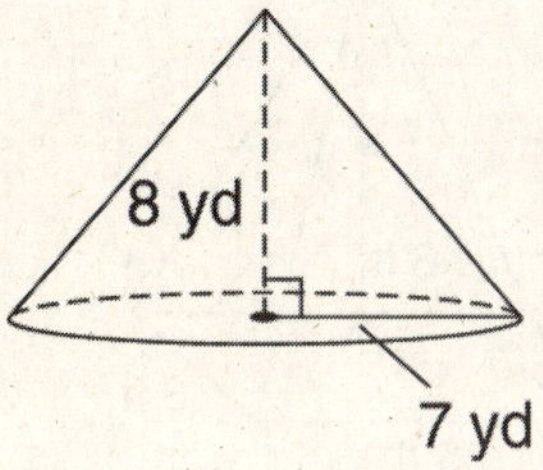

____

5.

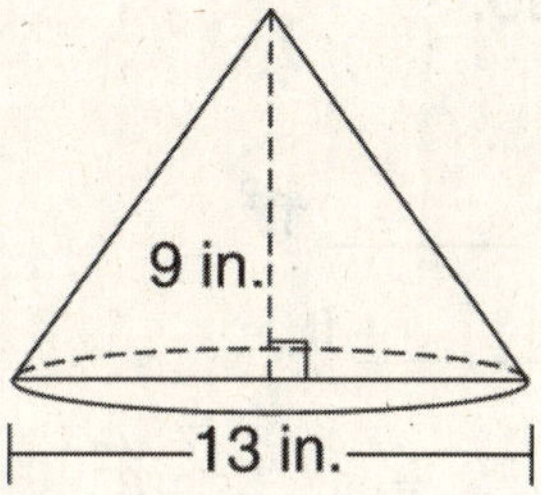

____

6.

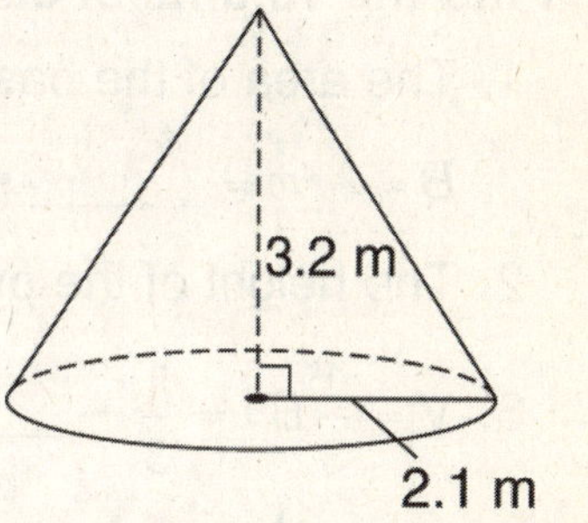

____

Find the volume of each solid to the nearest tenth. Use 3.14 for π.

7. rectangular pyramid 7 ft by 3 ft by 9 ft high ________________

8. cone with radius 8 m and height 11 m ________________

9. square pyramid with a 4 ft base and height 7 ft ________________

10. cone with diameter 15 in. and height 20 in. ________________

11. cone with diameter 11 in. and height of 6 in. ________________

12. triangular pyramid with height 10 cm and a base that is a right triangle with legs of 3 cm and 4 cm ________________

Name ______________________ Date ______________ Class ______________

LESSON **10-3**

Review for Mastery

Volume of Pyramids and Cones

The volume of a prism with base area B and height h is $V = Bh$.

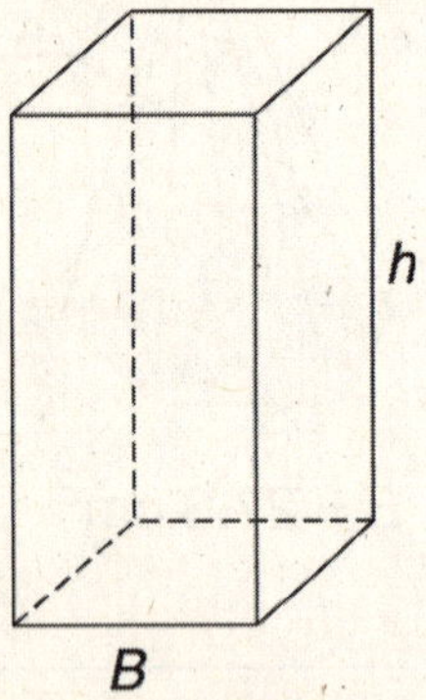

A pyramid with the same base B and height h has the volume $V = \frac{1}{3}Bh$.

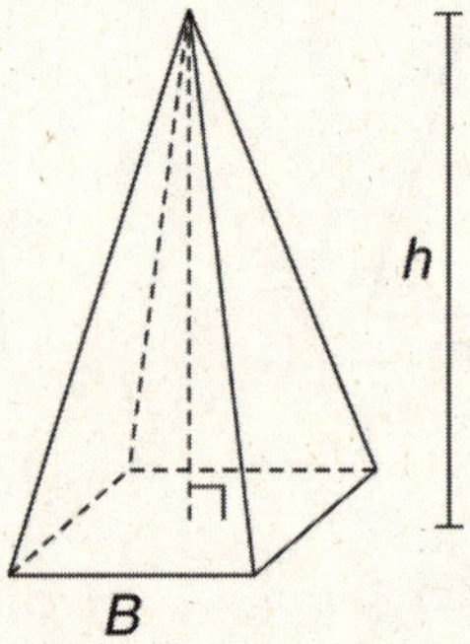

Find the volume of each pyramid.

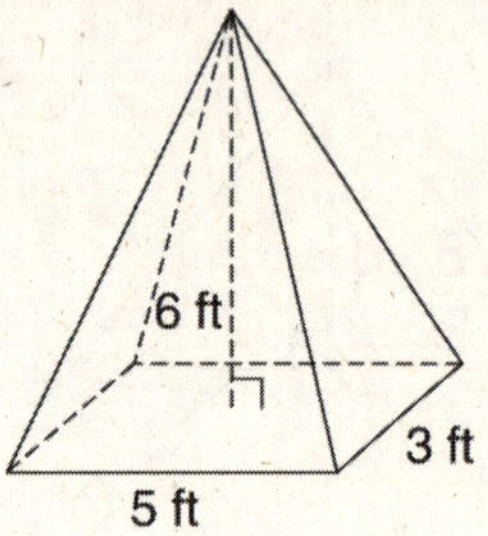

1. The area of the base is

   $B = \ell \bullet \omega =$ ______ • ______ = ______ ft^2

2. The height of the pyramid is ______ ft.

3. $V = \frac{1}{3}Bh = \frac{1}{3}$ • ______ • ______ = ______ ft^3

4. ______________

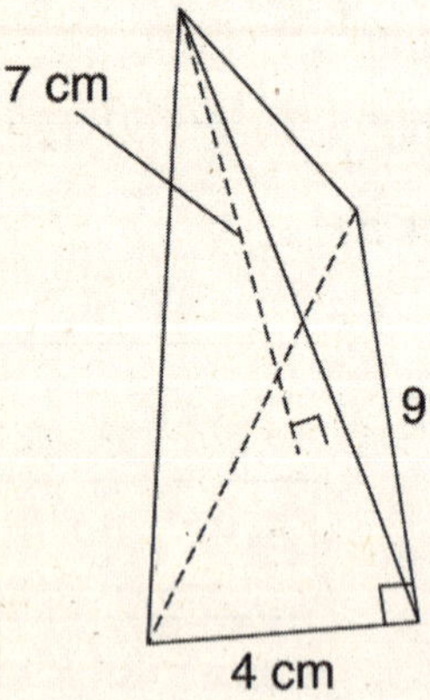

5. ______________

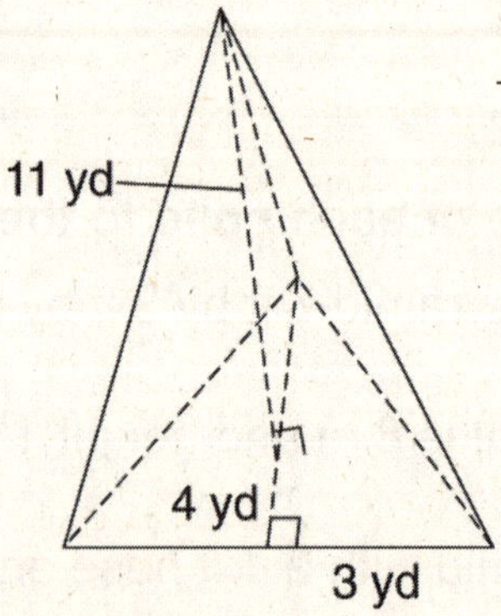

6. ______________

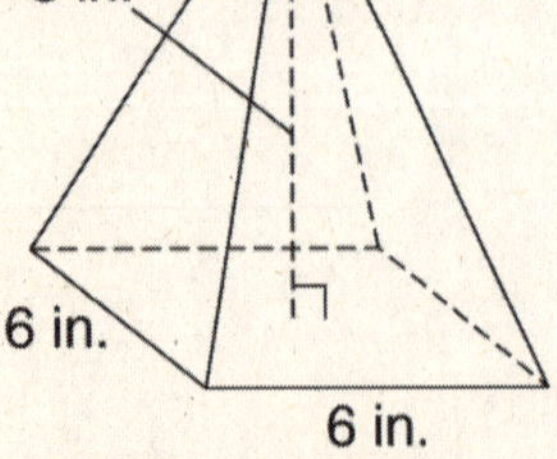

7. ______________

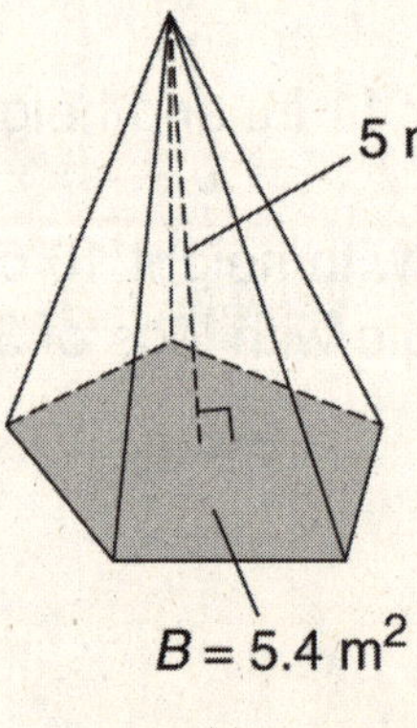

Name ______________________ Date ______________ Class ______________

LESSON 10-3

Review for Mastery

Volume of Pyramids and Cones (continued)

You can find the volume of a cone by using the formula $V = \frac{1}{3}Bh$.

Find the volume of each cone. Round to the nearest whole number for the final answer.

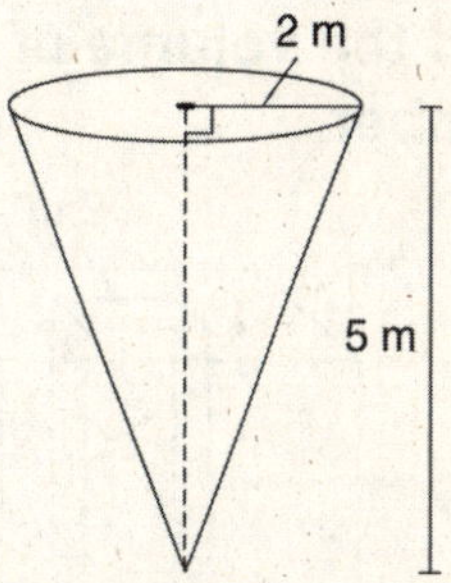

8. The area of the base is

$B = \pi r^2 \approx 3.14 \cdot$ ______$^2 \approx$ ______ m^2

9. $V = \frac{1}{3}Bh \approx \frac{1}{3} \cdot$ ______ $\cdot$ ______

$\approx \frac{1}{3} \cdot$ ______ $\approx$ ______ m^3

10.

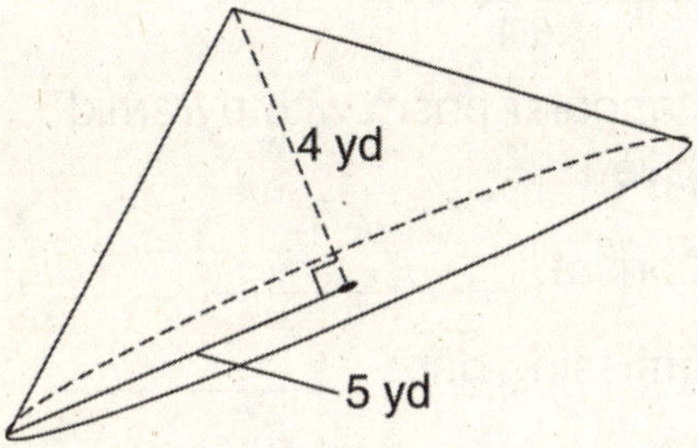

$B = \pi r^2$

$B \approx 3.14 \cdot$ ______^{2}

$\approx 3.14 \cdot$ ______ $\approx$ ______ yd^2

$V = \frac{1}{3}Bh$

$V \approx \frac{1}{3} \cdot$ ______ $\cdot$ ______

$\approx \frac{1}{3} \cdot$ ______ $\approx$ ______ yd^3

11.

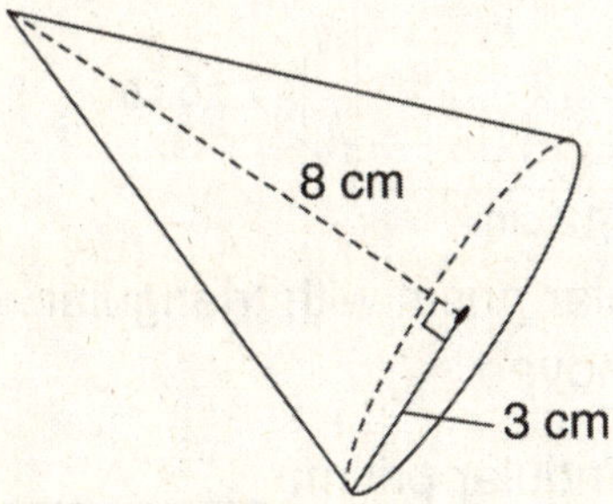

____

12.

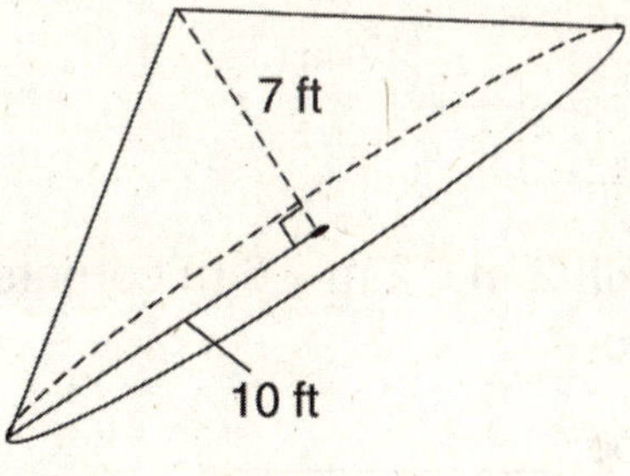

____

13.

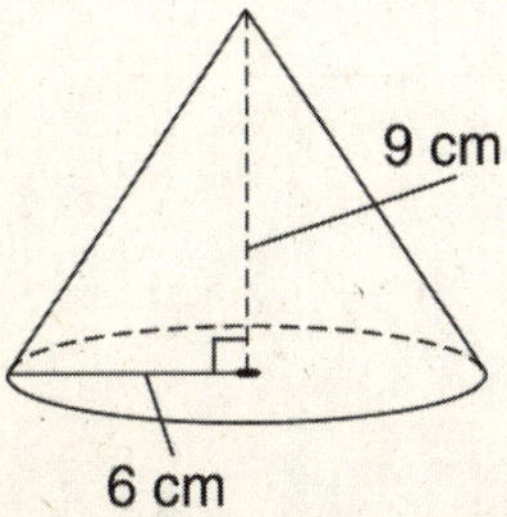

____

14.

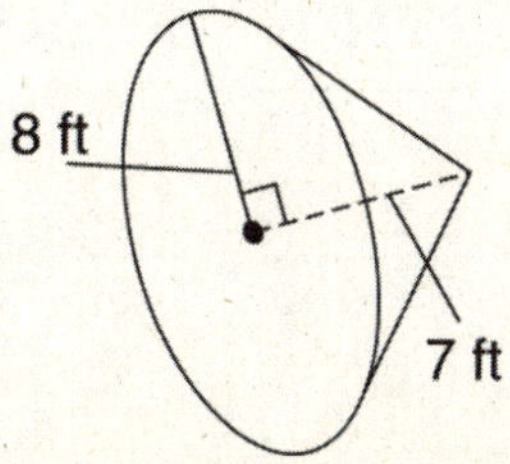

____

Name ______________________ Date ______________ Class ______________

LESSON 10-3

Challenge

Take It Out

Sometimes you can find the volume of an oddly shaped figure by first finding the volume of the entire solid and then subtracting a part that is missing.

Find the volume of each figure. Round to the nearest whole number.

1. 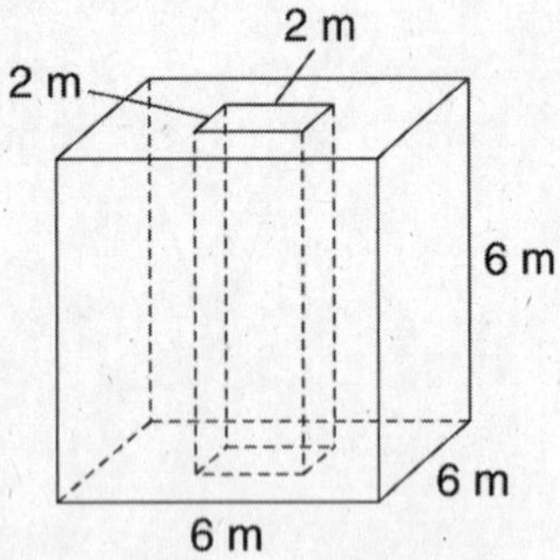

Cube with rectangular prism removed

V of cube: ____________

V of missing part: ____________

V of figure: ____________

2. 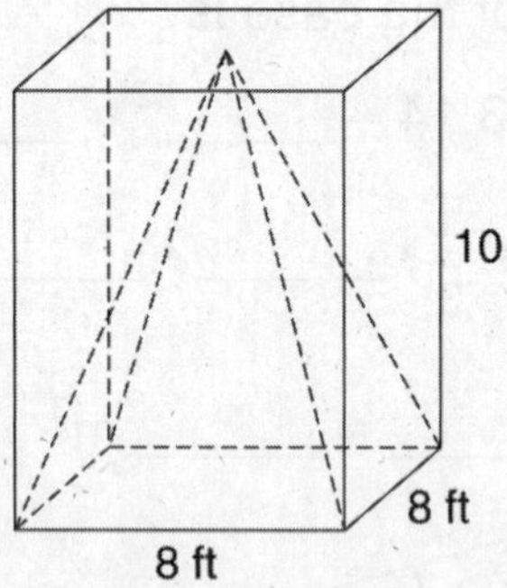

Rectangular prism with pyramid removed

V of prism: ____________

V of missing part: ____________

V of figure: ____________

3. 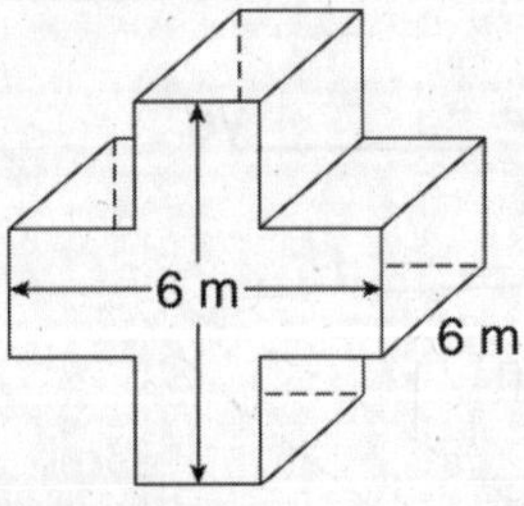

Cube with 2 m × 2 m × 6 m corners removed

V of cube: ____________

V of missing part: ____________

V of figure: ____________

4. 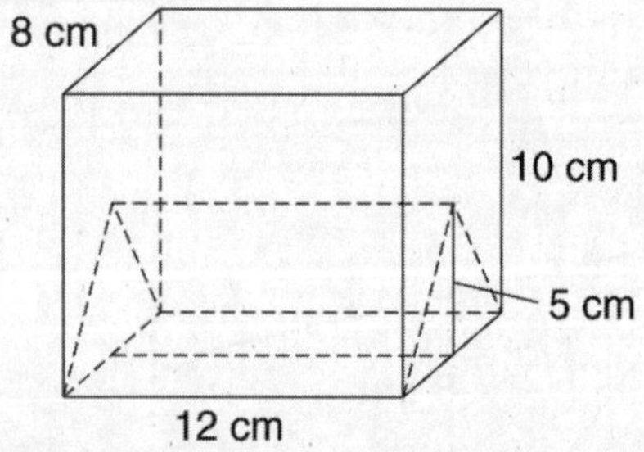

Rectangular prism with triangular prism removed

V of rectangular prism: ____________

V of missing part: ____________

V of figure: ____________

Name ______________________________ Date ______________ Class ______________

LESSON 10-3

Problem Solving

Volume of Pyramids and Cones

Write the correct answer.

1. Each of the cone-shaped cups near the water cooler has a radius of 3 centimeters and a height of 10 centimeters. If 1 cubic centimeter can hold 1 milliliter of liquid, how much water can each cup hold?

____

2. The Great Pyramid in Egypt has a square base that measures 751 feet on each side. The pyramid is 481 feet high. What is the volume of the Great Pyramid? Round your answer to the nearest cubic foot.

____

3. A waffle cone that holds ice cream is 15 centimeters high and has a diameter of 10 centimeters. What volume of ice cream can it hold if it is filled to the top?

____

4. The base of a rectangular prism is congruent to the base of a pyramid. The height of the pyramid is 3 times the height of the prism. Which figure has a greater volume? Explain.

____

____

____

Choose the letter of the correct answer.

5. A teepee that is shaped like a cone has a diameter of 12 feet and a height of 15 feet. What is the volume of the teepee?

A 565.2 ft^3 C 1,695.6 ft^3

B 706.5 ft^3 D 2,119.5 ft^3

6. The top of a 44-story office building is shaped like a pyramid. The base of the pyramid is a right triangle with the two legs measuring 73 feet and 78 feet. The pyramid is 35 feet high. What is the volume of the pyramid?

F 199,290 ft^3 H 41,756 ft^3

G 66,430 ft^3 J 33,215 ft^3

7. The diameter of a cone-shaped container is 4 inches. Its height is 6 inches. How much greater is the volume of a cylinder-shaped container with the same diameter and height?

A 50.24 in^3 C 100.98 in^3

B 75.36 in^3 D 200.96 in^3

8. A square pyramid mold for a candle has a base of 64 square centimeters and a height of 12 centimeters. How much greater is the volume of a rectangular prism mold with the same base and height?

F 64 cm^3 H 256 cm^3

G 96 cm^3 J 512 cm^3

Name ________________________ Date ______________ Class ______________

LESSON 10-3

Reading Strategies

Use a Graphic Organizer

This chart will help you compare pyramids and cones and the formulas for finding the volume of these figures.

Pyramids and Cones

Pyramid	**Volume of Pyramid**
• Base is a polygon. • Faces are triangles. • Vertex and base are at opposite ends.	Volume = $\frac{1}{3}B \cdot h$ (B = area of base)
Cone	**Volume of Cone**
• Base is a circle. • Has one curved surface. • Vertex is opposite the base.	Volume = $\frac{1}{3}\pi r^2 \cdot h$ (πr^2 = area of base)

Use the graphic organizer to answer each question.

1. Which figure has one curved surface? ____________
2. What forms the base of a pyramid? ____________
3. What forms the base of the cone? ____________
4. What is the formula for the area of a cone's base?

____

5. What shape are the faces of a pyramid?

____

6. Compare the volume formulas. How are they alike?

____

Name ______________________ Date ______________ Class ______________

LESSON 10-3

Puzzles, Twisters & Teasers

Loud and Clear!

Circle words from the list in the word search. Then find a word that answers the riddle. Circle it and write it on the line.

volume	pyramid	cone	sphere	container
prism	congruent	base	radius	cylinder

C	O	N	G	R	U	E	N	T	C	V
L	U	J	I	A	C	V	B	N	O	W
O	O	P	L	D	F	G	H	J	N	C
U	D	P	R	I	S	M	E	R	T	Y
D	M	N	B	U	A	S	D	F	A	L
P	C	B	A	S	E	T	Y	U	I	I
Y	W	D	R	T	H	U	I	J	N	N
R	W	S	C	D	E	C	O	N	E	D
A	T	H	U	Y	S	P	H	E	R	E
M	V	O	L	U	M	E	G	T	R	R
I	P	L	M	N	K	O	I	J	B	Y
D	Q	W	E	R	T	Y	U	I	I	O

How does the sky listen to music?

Through a ___________-speaker

CHAPTER 10

Family Letter

10B Surface Area

Dear Family,

The student will learn how to find the surface area of three-dimensional figures, such as prisms, cylinders, pyramids, and cones. **Surface area** is the sum of the areas of all of the surfaces of a three-dimensional figure expressed in square units. Some formulas for finding surface area are shown in the table below.

Figure	Words	Surface Area Formula
Prism	Twice the base area B plus the lateral area L. The lateral area is the base perimeter P times the height h.	$S = 2B + L$ or $S = 2B + Ph$
Cylinder	Twice the base area B plus the lateral area L. The lateral area is the base circumference $2\pi r$ times the height h.	$S = 2B + L$ or $S = 2\pi r^2 + 2\pi rh$
Pyramid	Base area B plus the lateral area L. The lateral area is one-half the base perimeter P times the slant height l.	$S = B + L$ or $S = B + \frac{1}{2}Pl$
Cone	Area of the circular base B plus the lateral area L. The lateral area is π times the radius r times the slant height l.	$S = B + L$ or $S = \pi r^2 + \pi rl$

Find the surface area of each figure to the nearest tenth. Use 3.14 for π.

A.

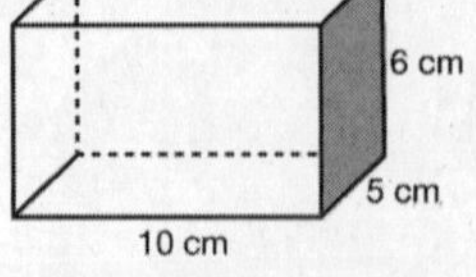

$S = 2B + Ph$ Use the formula.

$S = 2(10)(5) + (30)(6)$ Substitute. [$P = 10 + 5 + 10 + 5 = 30$]

$S = 100 + 180$ Multiply.

$S = 280$ Add.

The surface area of the prism is 280 cm^2.

Vocabulary

These are the math words we are learning:

lateral area (of a prism) the sum of the areas of the lateral faces

lateral face (of a prism) parallelograms that connect the bases

net an arrangement of two-dimensional figures that folds to form a three-dimensional figure

regular pyramid a pyramid whose base is a regular polygon and whose faces are congruent isosceles triangles

slant height of a cone the distance from the vertex to a point on the edge of the base

slant height of a pyramid the distance from the vertex opposite the base to the midpoint of an edge of the base

surface area the sum of the areas of all the surfaces of a figure

B.

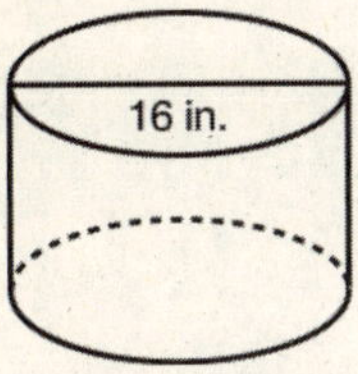

$S = 2\pi r^2 + 2\pi rh$	Use the formula.
$S = 2(\pi)(8)^2 + 2\pi(8)(9)$	Substitute; $r = 8$.
$S \approx 2(3.14)(8)^2 + 2(3.14)(8)(9)$	Substitute 3.14 for π.
$S \approx 2(3.14)(8)^2 + 2(3.14)(8)(9)$	Multiply.
$S \approx 401.92 + 452.16$	Add.
$S \approx 854.08$	Round to the nearest tenth.

The surface area of the cylinder is about 854.1 in^2.

The student will learn how to use scale factors to find the surface area and volume of similar figures.

The surface area of a cylinder is 100.5 in^2. What is the surface area of a similar cylinder that is larger by a scale factor of 3?

$S = 100.5 \cdot 3^2$	Multiply the surface area of the smaller cylinder by the square of the scale factor.
$S = 100.5 \cdot 9$	Evaluate the power.
$S = 904.5 \text{ in}^2$	

The surface area of the larger cylinder is 904.5 in^2.

Discuss with the student real-world applications of the concepts studied in this chapter.

Sincerely,

Name ______________________ Date ______________ Class ______________

CHAPTER 10

At-Home Practice

10B Surface Area

Find the surface area of each figure. Round to the nearest tenth, if necessary. Use 3.14 for π.

1.

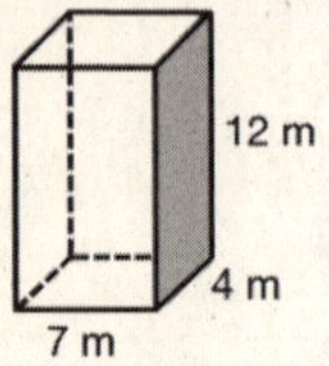

2.

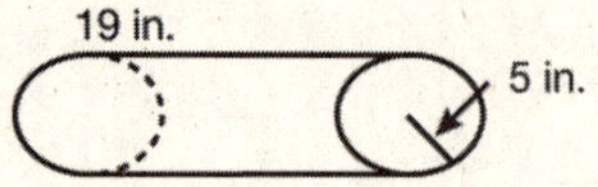

3.

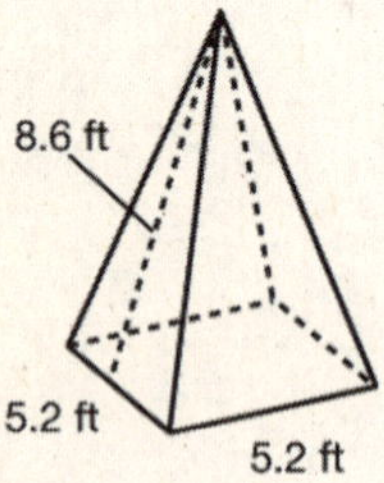

4.

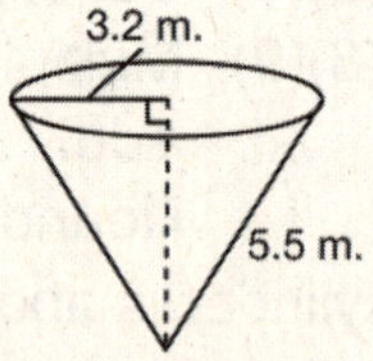

5. The surface area of a prism is 18 ft^2. What is the surface area of a similar prism that is larger by a scale factor of 4?

6. The volume of a cylinder is about 480 cm^3. What is the volume, to the nearest tenth, of a similar cylinder that is smaller by a scale factor of $\frac{1}{4}$?

Answers: 1. 320 m^2 **2.** 753.6 in^2 **3.** 116.5 ft^2 **4.** 87.4 m^2 **5.** 288 ft^2 **6.** 7.5 cm^3

Name ______________________ Date ____________ Class ____________

CHAPTER 10

Family Fun

Shape It up!

Directions

- Below are nets of different objects.
- Cut out the shapes using scissors.
- Piece them together and see what 3-dimensional shapes you can make.

Name ______________________ Date ______________ Class ______________

LESSON 10-4

Practice A

Surface Area of Prisms and Cylinders

Find the surface area of each prism. Choose the letter for the best answer.

1.

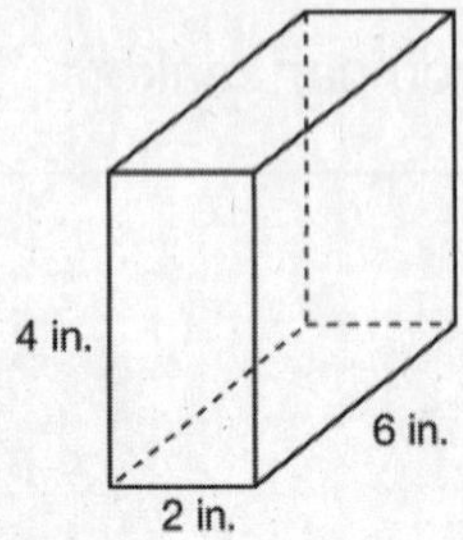

A 36 in^2 C 48 in^2

B 44 in^2 D 88 in^2

2.

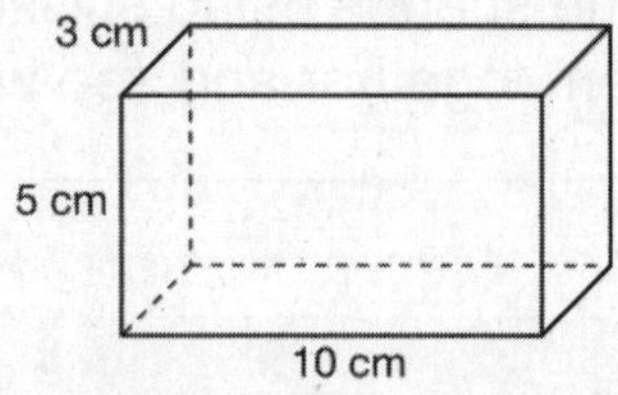

F 95 cm^2 H 190 cm^2

G 100 cm^2 J 380 cm^2

Find the surface area of each cylinder to the nearest whole number. Choose the letter for the best answer.

3.

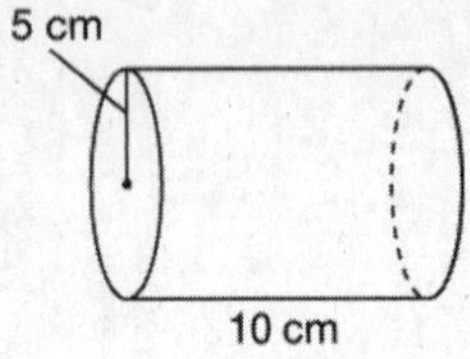

A 157 cm^2 C 471 cm^2

B 314 cm^2 D 942 cm^2

4.

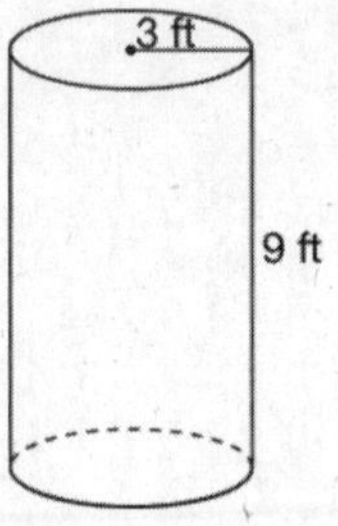

F 226 ft^2 H 113 ft^2

G 170 ft^2 J 57 ft^2

5. Dexter is making a sculpture with the dimensions shown. What is the surface area of the sculpture? Use 3.14 for π.

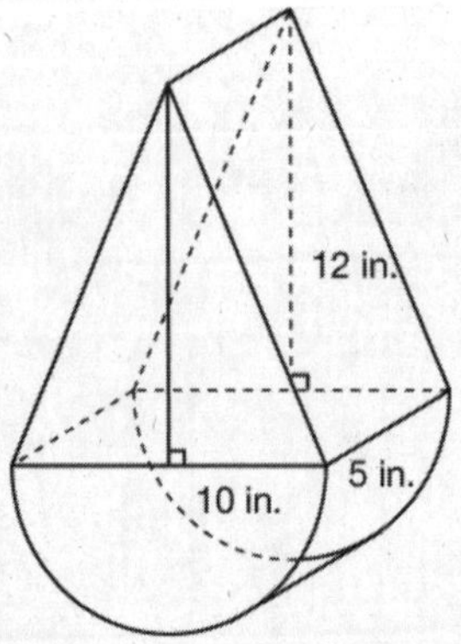

____

Name ______________________ Date ______________ Class ______________

LESSON 10-4

Practice B

Surface Area of Prisms and Cylinders

Find the surface area of each prism to the nearest tenth.

1.

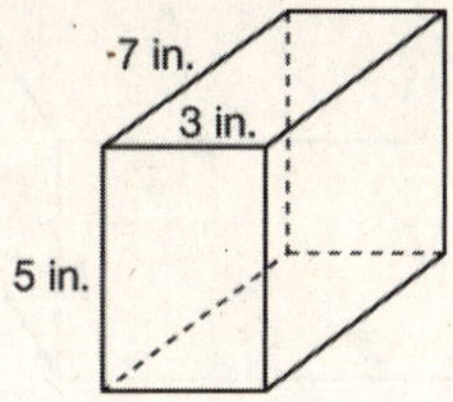

____

2.

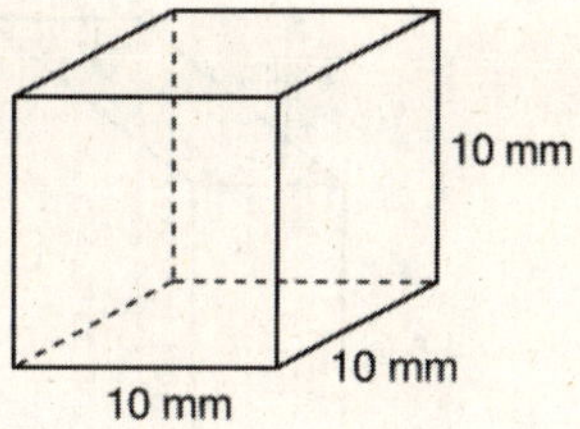

____

Find the surface area of each cylinder to the nearest tenth.

3.

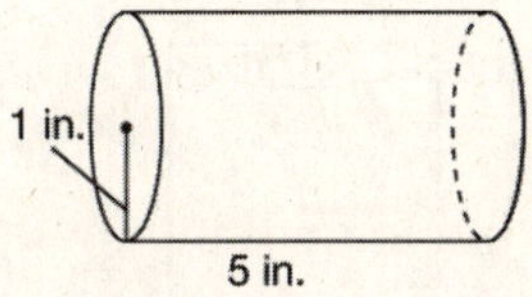

____

4.

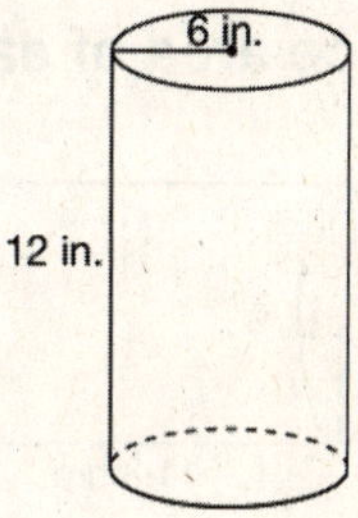

____

5. Trina is designing a tent with the dimensions shown. What is the surface area of the tent? Round to the nearest tenth.

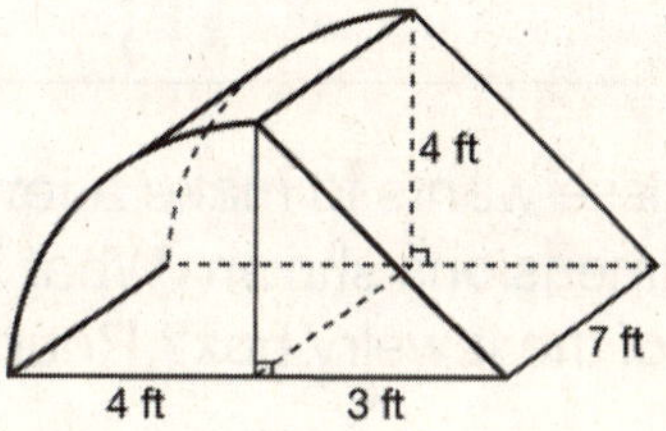

____

Name ______________________ Date ______________ Class ______________

LESSON 10-4

Practice C

Surface Area of Prisms and Cylinders

Find the surface area of each prism to the nearest tenth.

1.

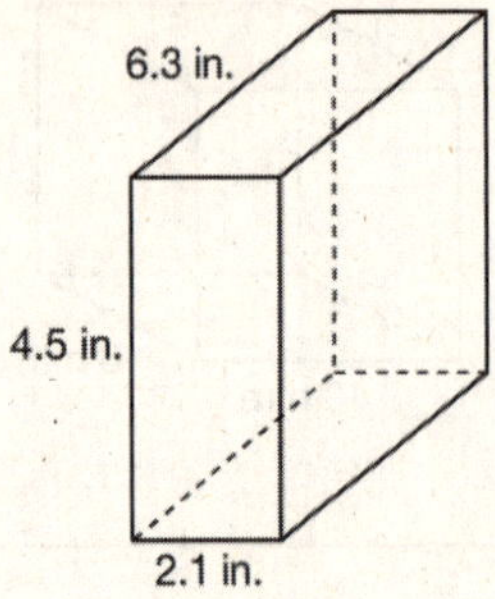

____

2.

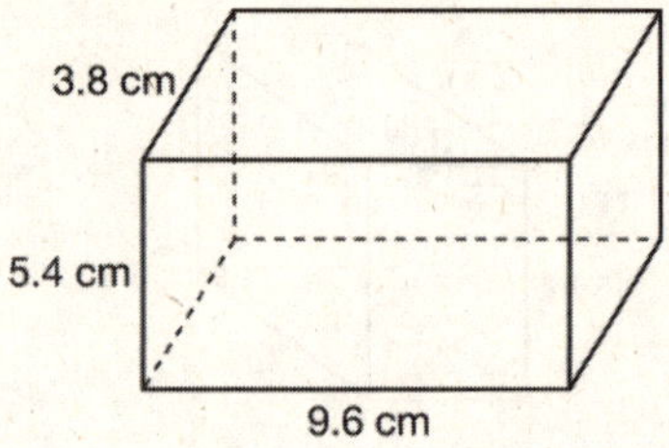

____

Find the surface area of each cylinder to the nearest tenth.

3.

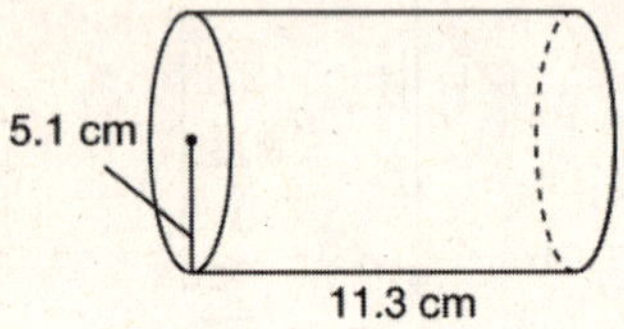

____

4.

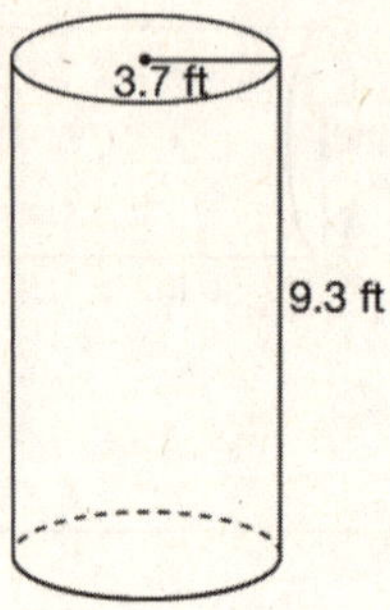

____

5. Amarisse wants to make a jewelry box with the dimensions shown. What is the surface area of the jewelry box? Round to the nearest tenth.

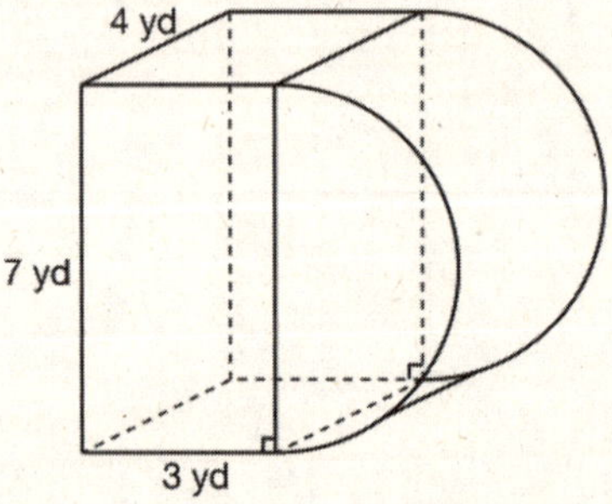

____

Name ______________________ Date ____________ Class ____________

LESSON 10-4

Review for Mastery

Surface Area of Prisms and Cylinders

The surface area of a three-dimensional figure is the combined areas of the faces. You can find the surface area of a prism by drawing a **net** of the flattened figure.

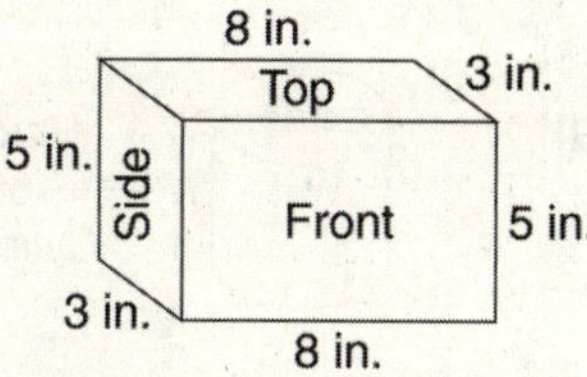

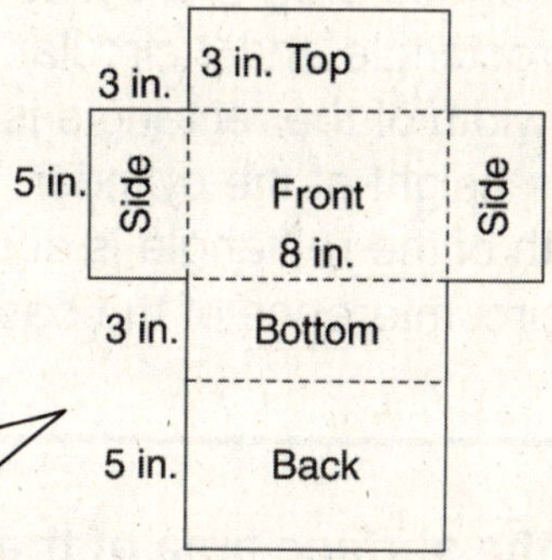

Notice that the front and back faces are the same. The side faces are the same. The top and bottom faces are the same.

Find the surface area of the prism formed by the net above.

1. Find the area of the front face: *A* = ______ • ______ = ______ in^2.

   The area of the front and back faces is 2 • ______ = ______ in^2.

2. Find the area of the side face: *A* = ______ • ______ = ______ in^2.

   The area of the 2 side faces is 2 • ______ = ______ in^2.

3. Find the area of the top face: *A* = ______ • ______ = ______ in^2.

   The area of the top and bottom faces is 2 • ______ = ______ in^2.

4. Combine the areas of the faces: ______ + ______ + ______ = ______ in^2.

   The surface area of the prism is ______ in^2.

Find the surface area of the prism formed by each net.

5.

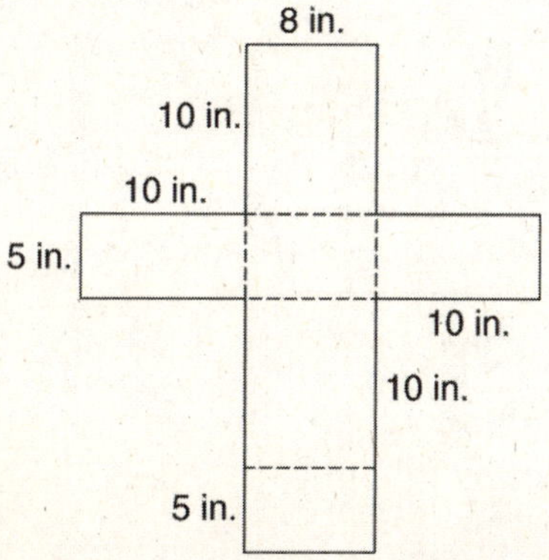

____

6.

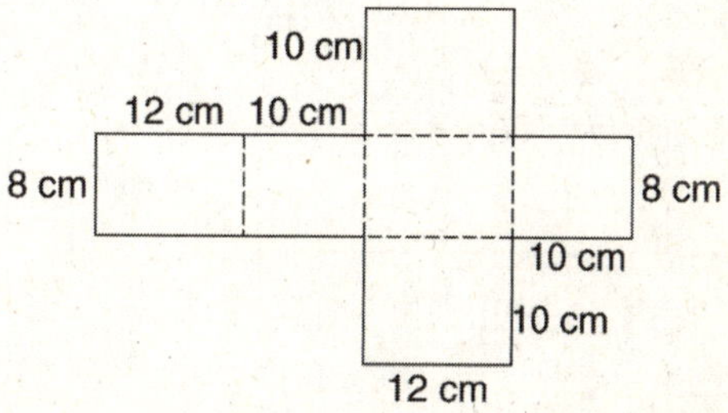

____

Name ______________________ Date ______________ Class ______________

LESSON 10-4

Review for Mastery

Surface Area of Prisms and Cylinders (continued)

The surface area of a cylinder consists of a rectangle and 2 circular bases. The width of the rectangle is equal to the height of the cylinder. The length of the rectangle is equal to the circumference of the bases.

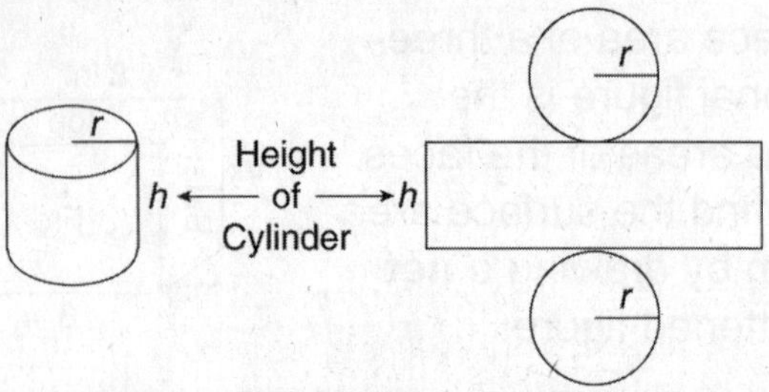

Find the surface area of the cylinder formed by the net. Round to the nearest whole number for the final answer.

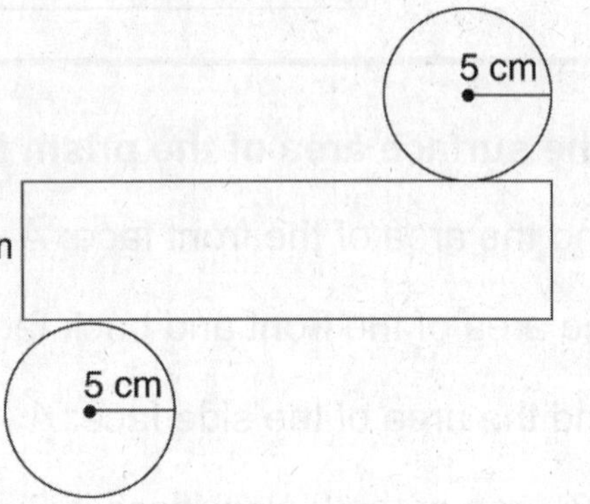

7. Find the area of each circular base.

   $A = \pi r^2 = 3.14 \cdot$ ______ = ______ cm^2

8. Find the area of both circular bases.

   $A = 2 \cdot$ ______ = ______ cm^2

9. Find the area of the rectangle.

   $A = \ell \cdot w$

   $A = 2\pi r \cdot h$

   $A = 2 \cdot 3.14 \cdot$ ______ $\cdot$ ______

   $A =$ ______ cm^2

10. Combine the areas of the circular bases and the area of the rectangle to find the surface area, S, of the cylinder.

   $S =$ ______ + ______ = ______ cm^2

Find the surface area of the cylinder formed by each net. Round to the nearest whole number. Use 3.14 for π.

11.

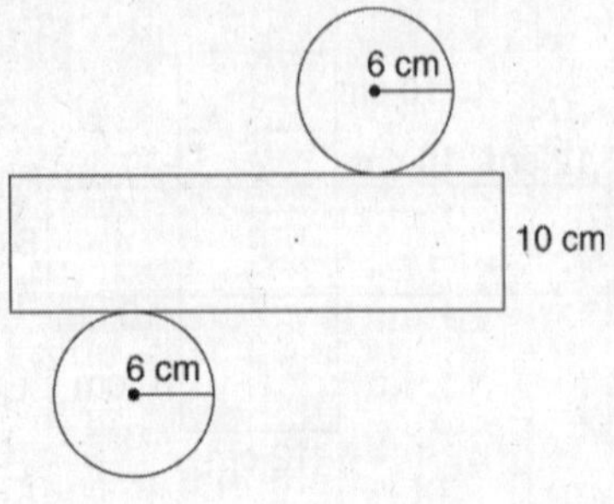

____

12.

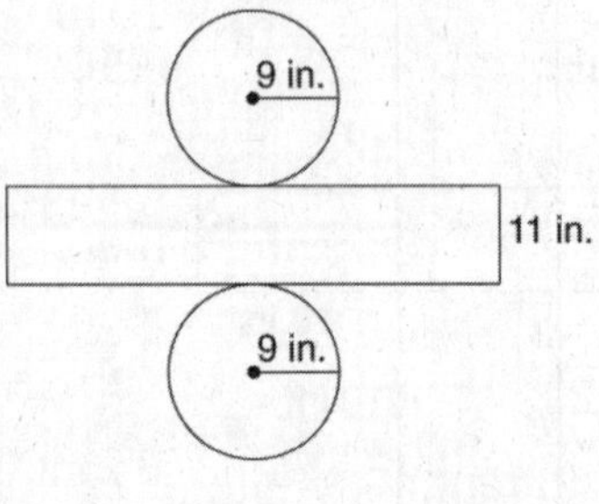

____

Name __________________ Date __________ Class __________

LESSON 10-4

Challenge

Saving Trees

Boxes with different dimensions can have the same volume. Marcus Manufacturing uses boxes to ship office supplies to stores. The less cardboard used to make a box, the less the box costs.

In each set, the boxes have equal volume. Find the missing dimension, if necessary. Write the surface area below each box. Then circle the box in each set that uses the least amount of cardboard.

1.

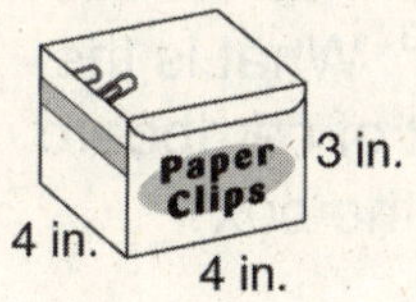

surface area = __________

width = __________

surface area = __________

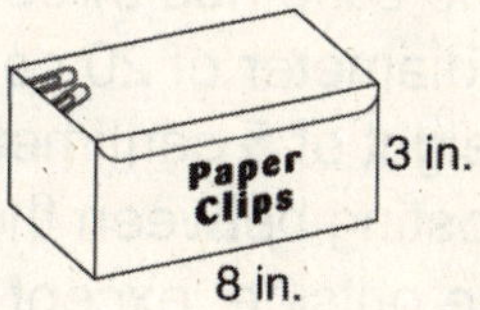

width = __________

surface area = __________

2.

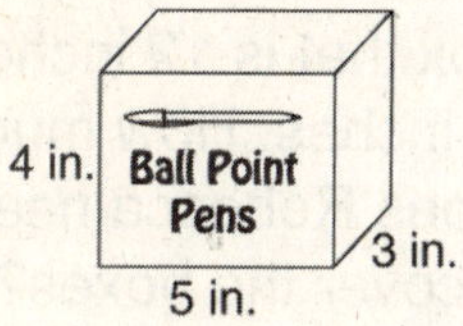

surface area = __________

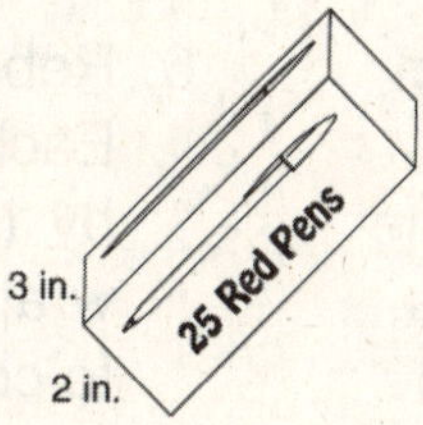

width = __________

surface area = __________

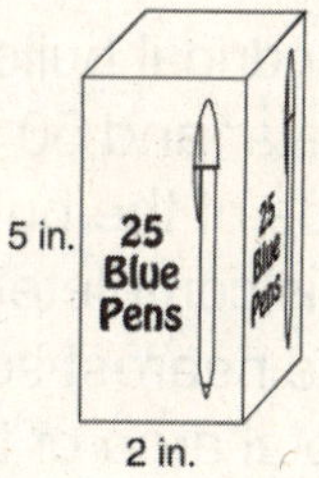

width = __________

surface area = __________

3.

width = __________

surface area = __________

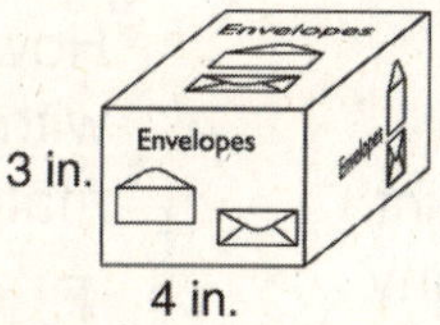

width = __________

surface area = __________

width = __________

surface area = __________

Name ______________________ Date ______________ Class ______________

LESSON 10-4

Problem Solving

Surface Area of Prisms and Cylinders

Write the correct answer.

1. A can of peas is 3 inches in diameter and 4.5 inches tall. What is the area of the label used around the can?

____

2. How much wrapping paper do you need to completely cover a rectangular box that is 20 inches by 18 inches by 2 inches?

____

3. Jan puts frosting on a circular cake. The cake has three layers, each with a diameter of 20 centimeters and a height of 5 centimeters. Jan puts frosting between the layers and on the outside, except for the bottom. What is the area that Jan frosts? Round to the nearest square centimeter.

____

4. A cardboard storage carton has a length of 3 feet, a width of 2 feet, and a volume of 12 ft^3. What is the minimum amount of cardboard needed to make the box?

____

Choose the letter of the correct answer.

5. A cylindrical building is 30 meters in diameter and 50 meters high. The outside of the building, excluding the roof, is completely covered in glass. To the nearest square foot, what is the total area of the glass?

   A 1,413 m^3 C 6,113 m^3

   B 4,710 m^3 D 9,420 m^3

6. Rebecca gives gifts to 12 employees. Each gift is in a box that is 12 inches by 10 inches by 3 inches. How much wrapping paper does Rebecca need to completely the cover the boxes?

   F 360 in^2 H 4,464 in^2

   G 1,720 in^2 J 14,400 in^2

7. A cylinder-shaped sculpture is 24 meters high with a diameter of 6.8 meters. An artist plans to spray-paint the entire surface with silver paint. If one can of spray paint covers 50 square meters, how many cans does the artist need to paint the sculpture?

   A 51 cans C 12 cans

   B 22 cans D 10 cans

8. A rectangular sofa cushion is 36 inches by 30 inches by 20 inches. How many cushions can be covered with 38,400 square inches of material?

   F 6 cushions H 39 cushions

   G 8 cushions J 10 cushions

Name ______________________ Date ______________ Class ______________

LESSON 10-4

Reading Strategies

Analyze Information

If you unfold a three-dimensional figure and lay it flat, you have made a net.

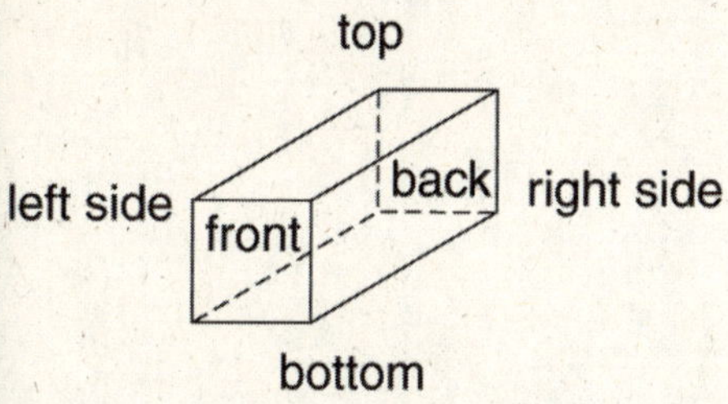

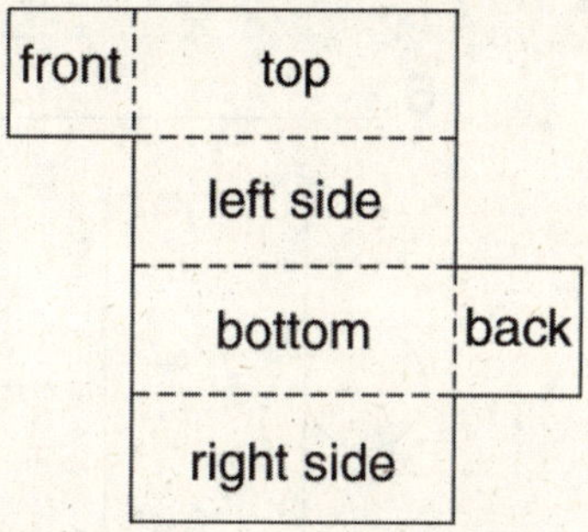

A **net** is a two-dimensional shape that lets you picture all the surfaces of a three-dimensional figure. A net helps you see how much surface a three-dimensional figure covers.

The **surface area** of a three-dimensional figure is the total area of all its surfaces.

If you analyze the net of a rectangular prism, you notice there are six faces. Each face pairs up with another, congruent face:

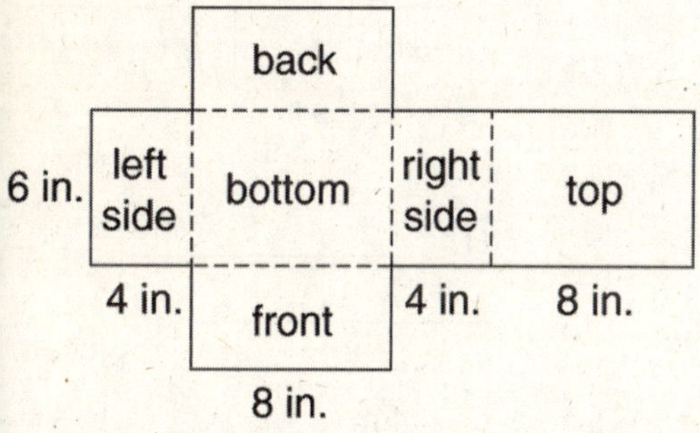

↔

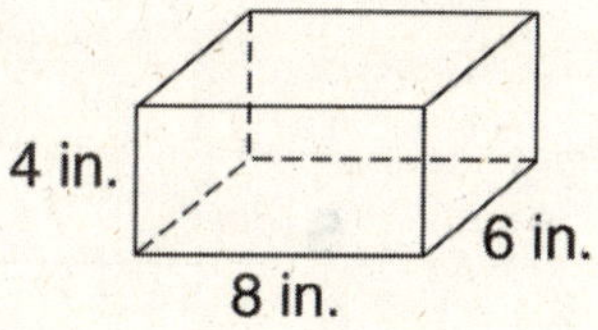

To find the surface area of a rectangular prism, find the sum of the areas of the six faces (rectangles). *Hint:* $A = \ell w$.

Answer the following to find the surface area of the rectangular prism.

1. What is the area of the bottom rectangle? ______________
2. What is the area of the top rectangle? ______________
3. What is the area of the back rectangle? ______________
4. What is the area of the front rectangle? ______________
5. What is the area of the right rectangle? ______________
6. What is the area of the left rectangle? ______________
7. What is the surface area of the rectangular prism? ______________

Name ______________________ Date ______________ Class ______________

LESSON 10-4

Puzzles, Twisters & Teasers

Bee a Math Marvel!

Find the surface area of each figure below. Round to the nearest tenth. Use your answers to solve the riddle.

L ______________

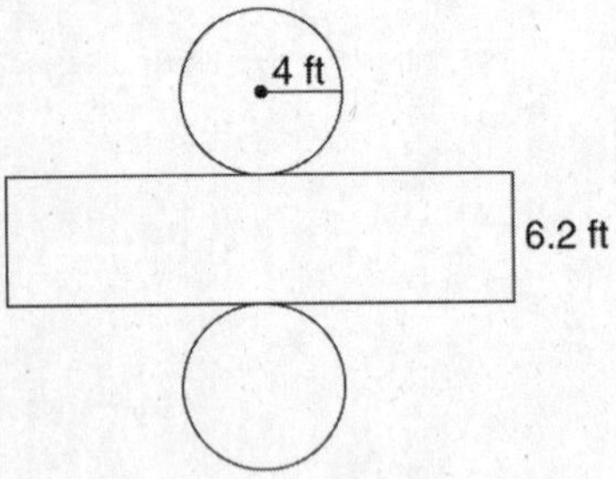

B ______________

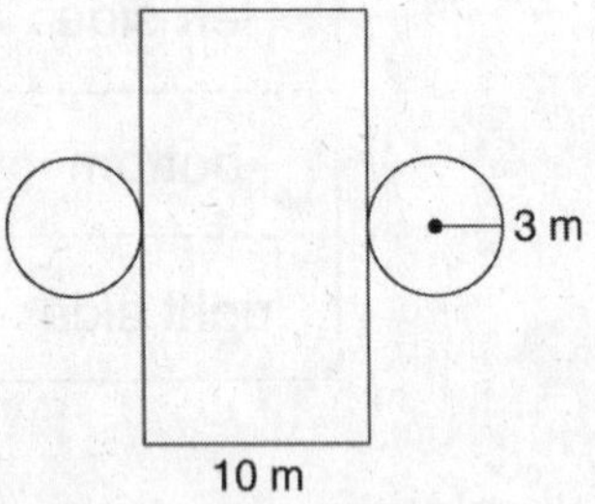

A ______________

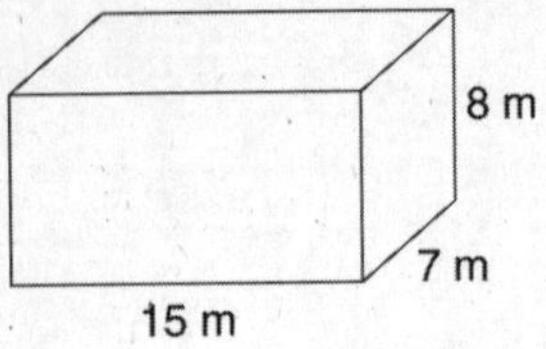

H ______________

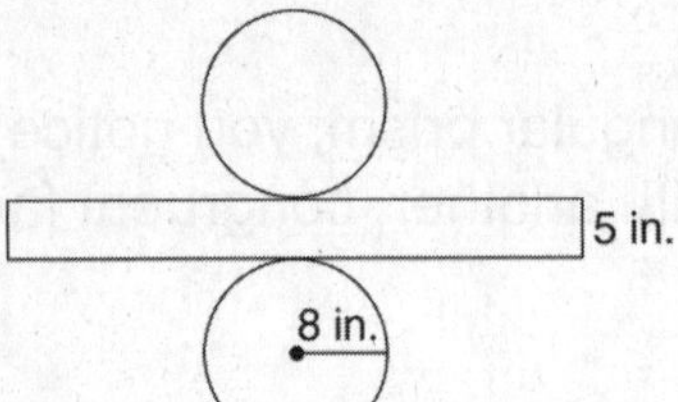

T ______________

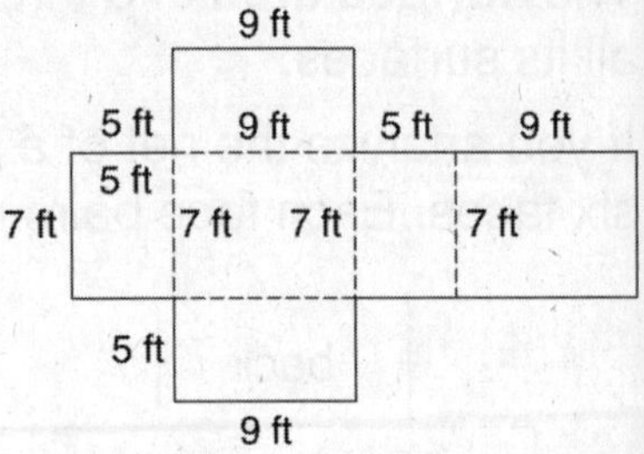

O ______________

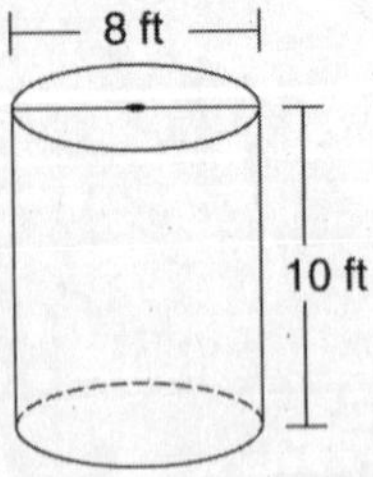

S ______________

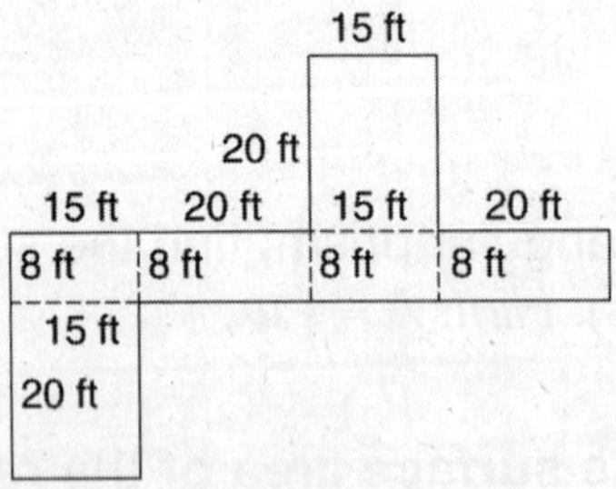

I ______________

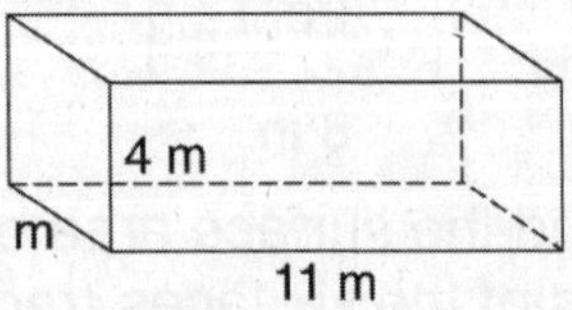

Why couldn't the bee go to the dance?

___	___	W ___	___	___	M ___	___	___	___	___	___	___	___
268	286	562	1,160	562	351.7	286	653.1	244.9	562	256.2	256.2	

Name ______________________ Date ______________ Class ______________

LESSON 10-5

Practice A

Surface Area of Pyramids and Cones

Find the surface area of each pyramid.

1. 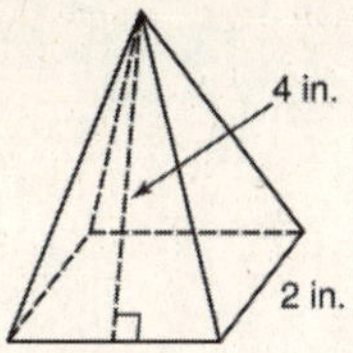

____

2. 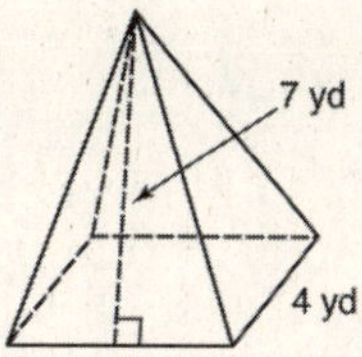

____

3. 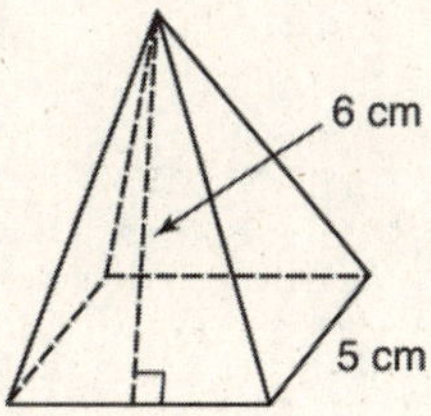

____

4. 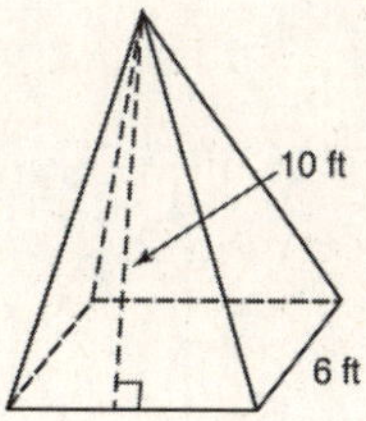

____

5. 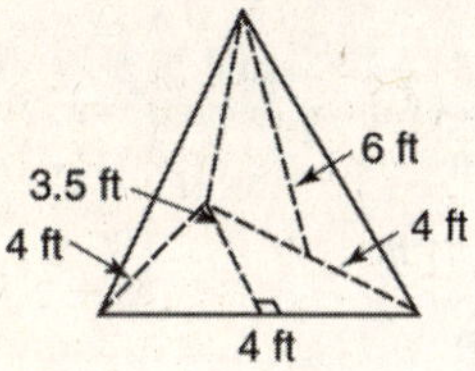

____

6. 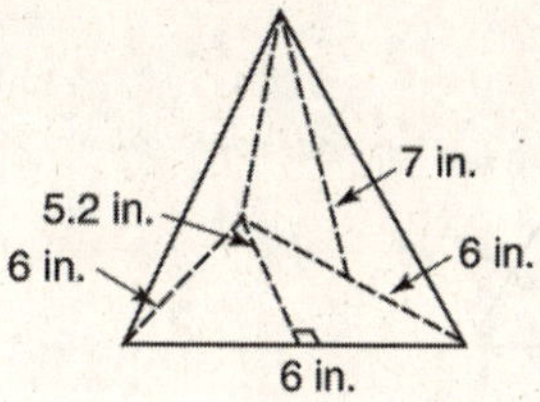

____

7. 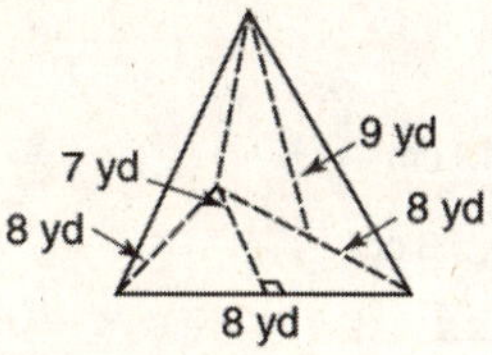

____

8. 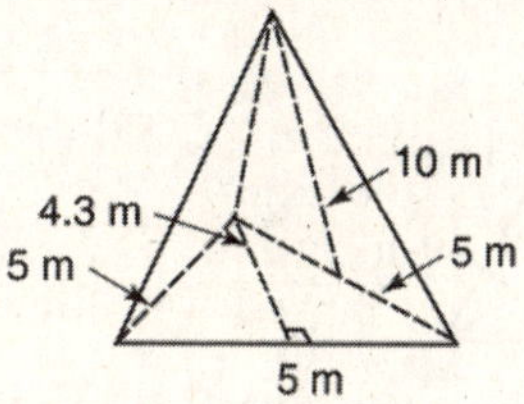

____

Find the surface area of each cone. Use 3.14 for π.

9. 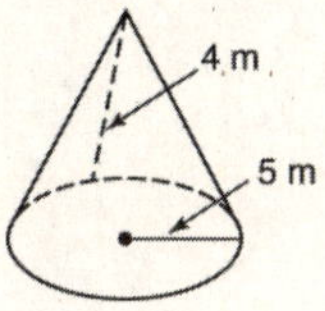

____

10. 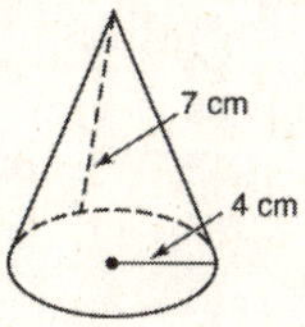

____

11. 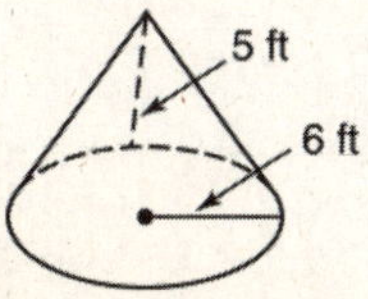

____

12. 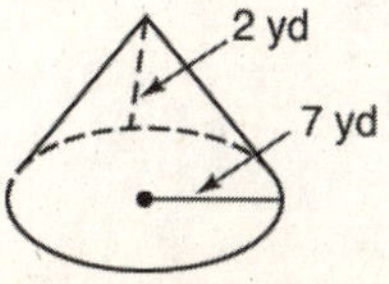

____

Name ______________________ Date ______________ Class ______________

LESSON 10-5

Practice B

Surface Area of Pyramids and Cones

Find the surface area of each pyramid.

1.
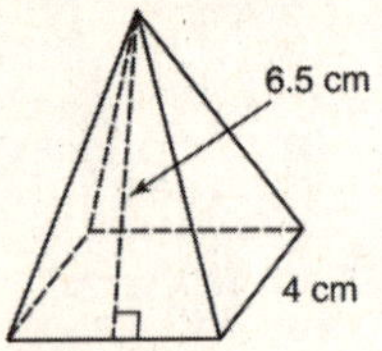

____

2.
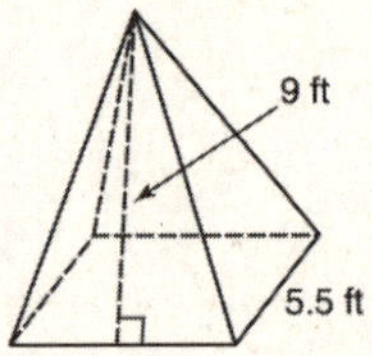

____

3.
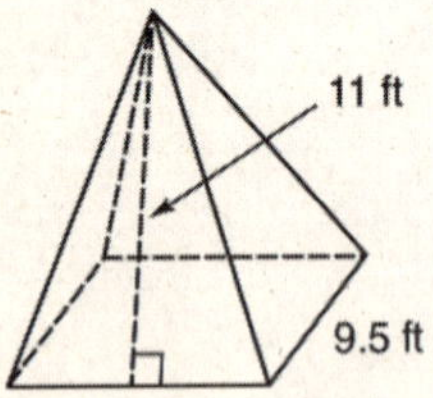

____

4.
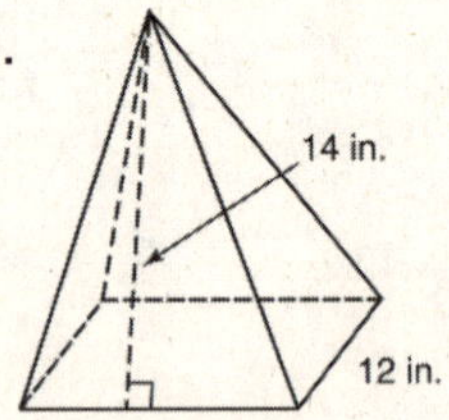

____

5.
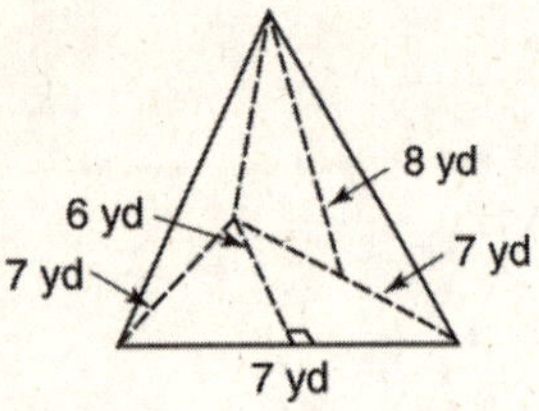

____

6.
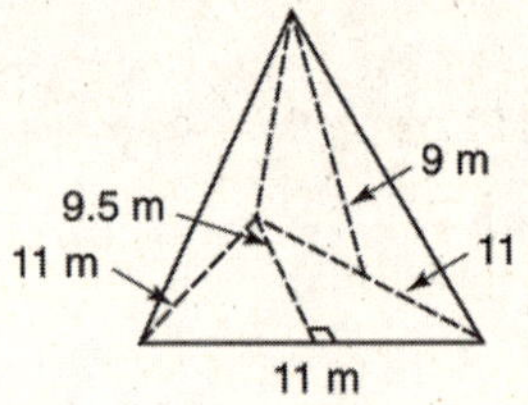

____

7.
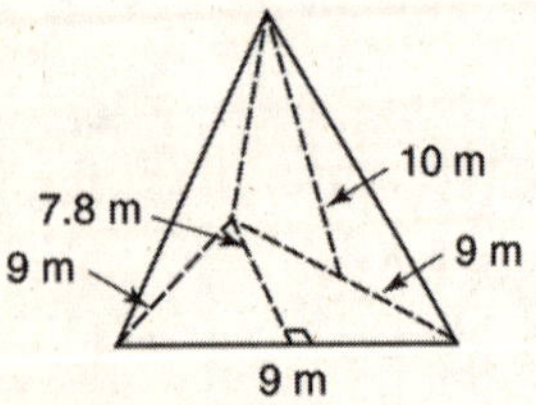

____

8.
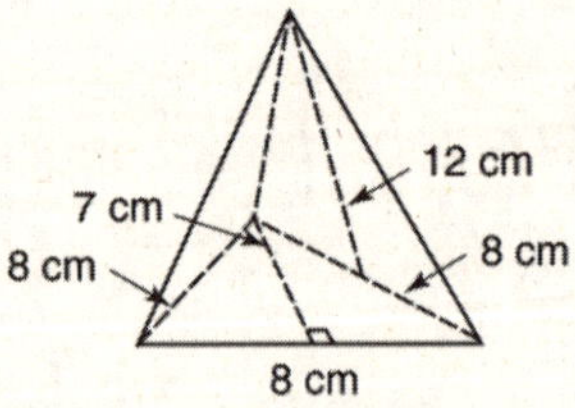

____

Find the surface area of each cone. Use 3.14 for π.

9.
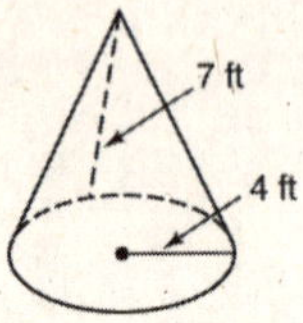

____

10.
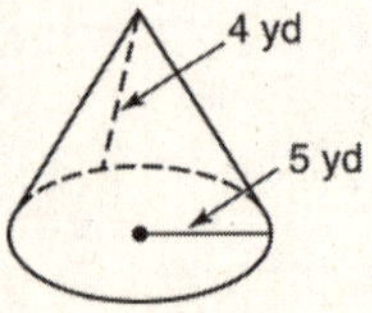

____

11.
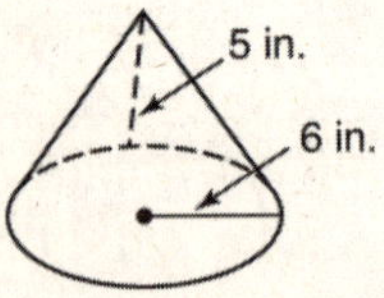

____

12.
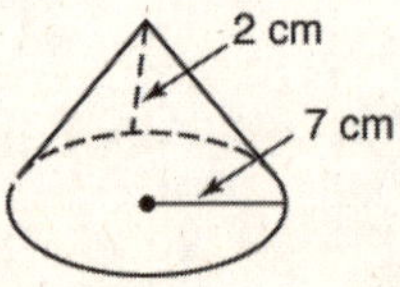

____

Name ______________________ Date ______________ Class ______________

LESSON 10-5

Practice C

Surface Area of Pyramids and Cones

Find the surface area of each pyramid.

1.
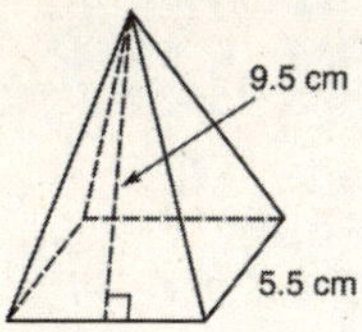

____

2.
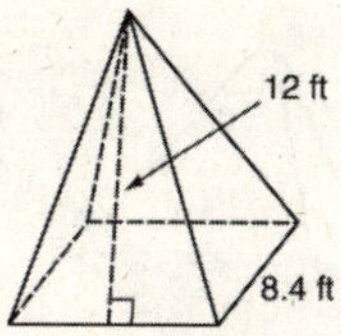

____

3.
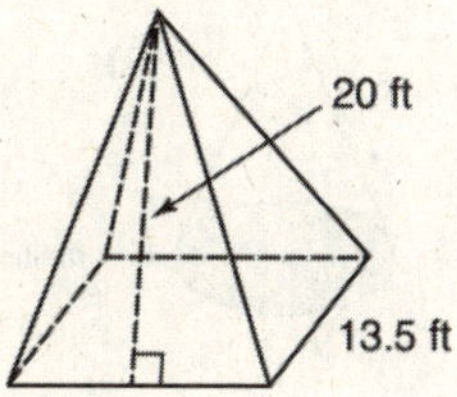

____

4.
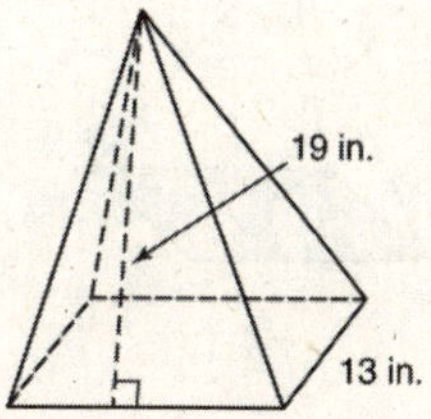

____

5.
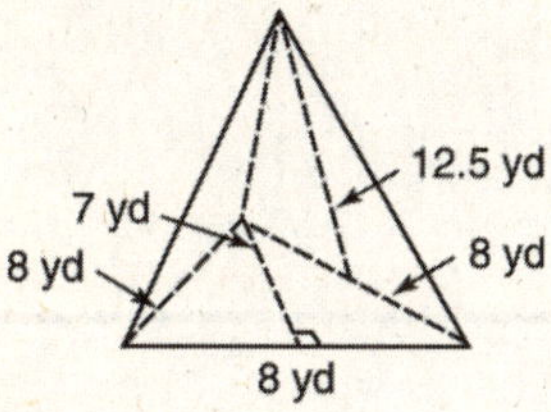

____

6.
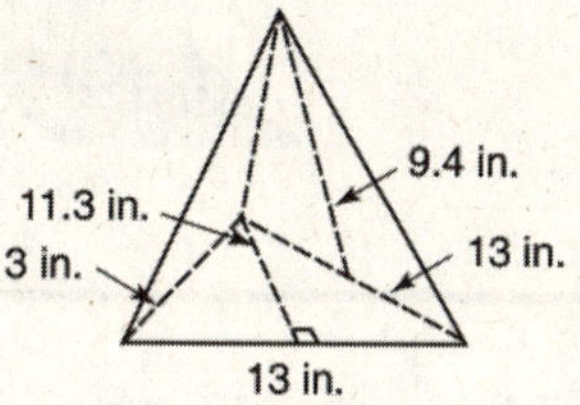

____

7.
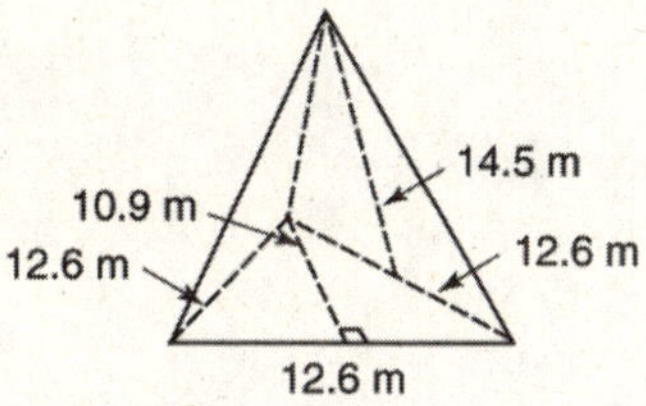

____

8.
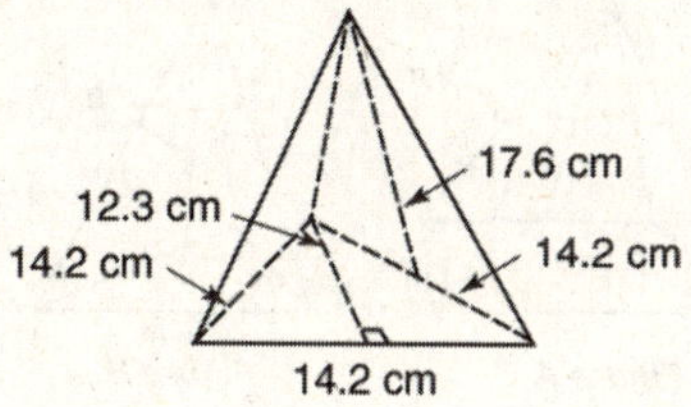

____

Find the surface area of each cone. Use 3.14 for π.

9.
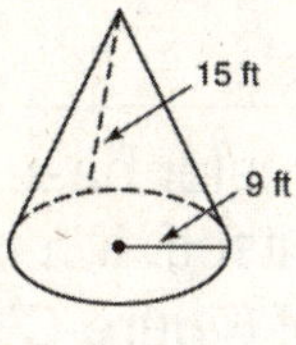

____

10.
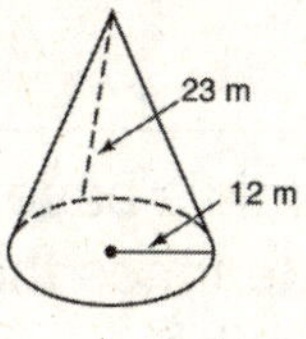

____

11.
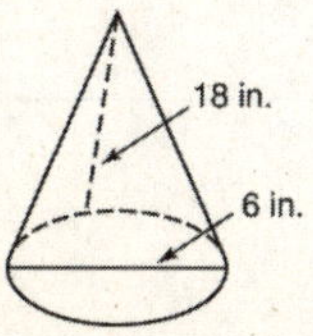

____

12.
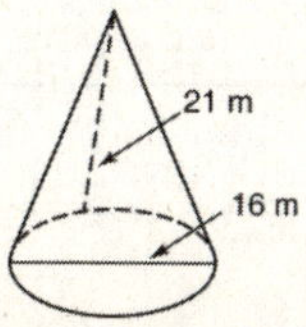

____

Name ______________________ Date ______________ Class ______________

LESSON 10-5

Review for Mastery

Surface Area of Pyramids and Cones

Finding the surface area of a pyramid or a cone involves knowing certain terms. These include:

- The **base area:**

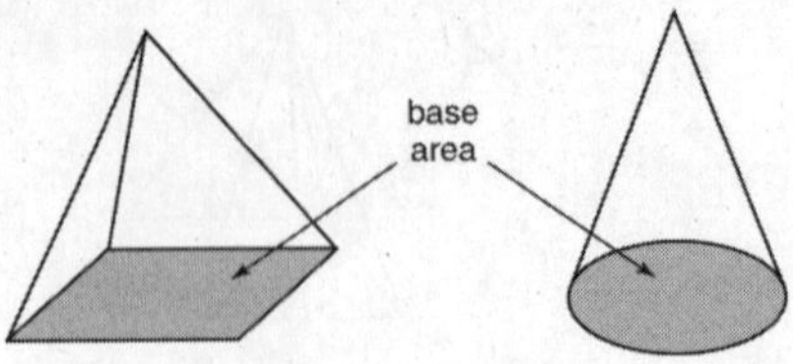

- The **base side:**

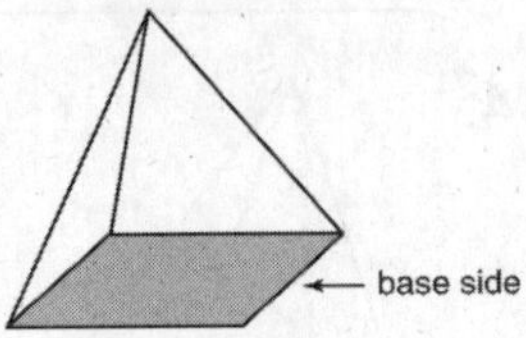

- The **radius:**

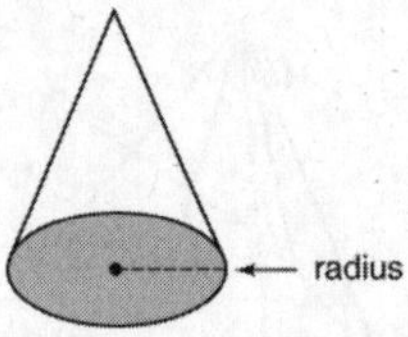

- The **slant height:**

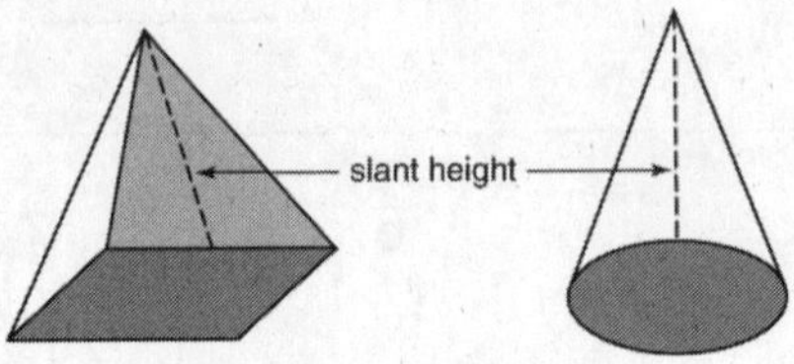

Use the drawings to answer the questions.

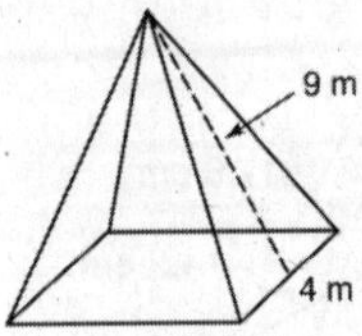

Figure A

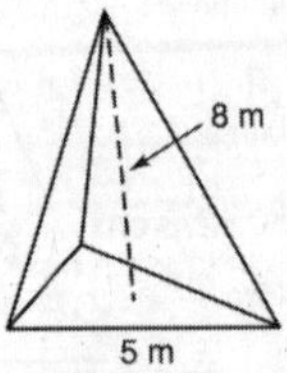

Figure B

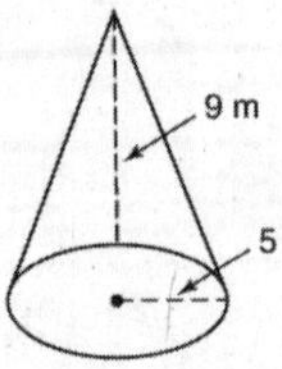

Figure C

1. How long is the base side of Figure B?

____

2. Which two figures have the same slant height?

____

3. The perimeter of a pyramid's base is the length of its base side times the number of sides. What is the perimeter of Figure A? Of Figure B?

____

____

4. The area of a cone's circular base is π times its radius times itself. If π is 3.14, what is the area of Figure C's base?

____

Name ______________________ Date ____________ Class ____________

LESSON 10-5

Challenge

Finding the Slant Height of a Cone

Below are the top and side views of the same cone:

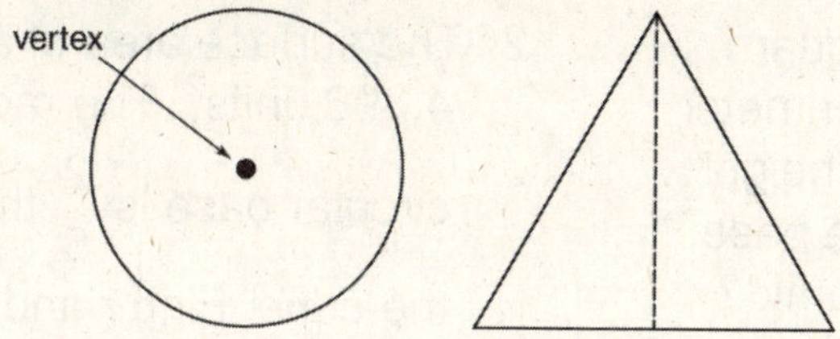

Notice that the vertex is perfectly centered over the base. The dotted line in the side view shows that if the vertex were extended downward, it would form a line perpendicular to the base and passing through its center. This line is the **altitude** of the cone, and it is used in determining the *slant height*. Here is the formula:

$$\ell = \sqrt{r^2 + H^2},$$

where r is the radius and H is the altitude.

The formula for the surface area of a cone excluding its base is, thus, $\pi r\sqrt{r^2 + H^2}$.

For Exercises 1–5, use $\pi = 3.14$.

1. What is the slant height of a cone whose base radius is 5 cm and whose altitude is 12 cm?

____

2. What is the slant height of a cone whose base radius is 4 cm and whose altitude is 3 cm?

____

3. How can you derive the formula for the surface area of a cone excluding its base (S_{-B})?

____

4. What is an application that might call for determining the surface of a cone but not the base?

____

5. What is the relationship between finding the slant height of a cone and the Pythagorean Theorem?

____

Name ______________________ Date ______________ Class ______________

LESSON 10-5

Problem Solving

Surface Area of Pyramids and Cones

Solve the problems.

1. A three-sided pyramid and square pyramid both have a base perimeter of 30 cm and the same slant height and surface area. What is the base height of the three-sided pyramid?

____

____

2. The surface area of a cone is exactly 4,396 units^2. The radius of the circular base is $\frac{2}{5}$ the slant height of the cone. Find r and ℓ. Use 3.14 for π.

____

____

3. The Ladrones Tower in Marbella, Spain, is a cone-shaped building with a base circumference of 12 m and a slant height of 14.7 m. Find the surface area of the structure. *Hint*: The formula $\frac{C}{d} = \pi$ expresses the relationship between the circumference and diameter of a circle.

____

____

4. The Transamerica Building in San Francisco is a square pyramid with a metal top. The metal top has a base perimeter of 43 m and a slant height of 65 m. How many square meters of aluminum were required for the top of the Transamerica Building?

____

____

____

At the right is a square pyramid designed by Chinese-American architect I.M. Pei. Use it to answer the questions. Circle the letter of the correct answer.

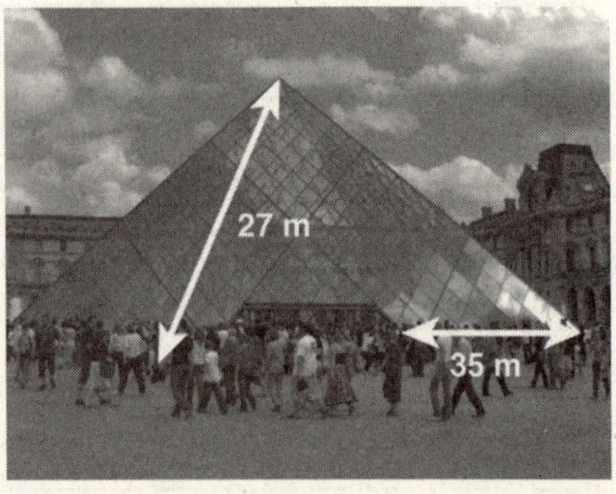

5. The pyramid's base area is

   A 35 m

   B 140 m^2

   C 1,225 m^2

   D 122 m

6. The surface area is of the glass walls is

   A 1,890 m^2

   B 2,112 m^2

   C 4,114 m^2

   D 6,028 m^2

Name ____________________ Date ____________ Class ____________

LESSON 10-5

Reading Strategies

Surface Area of Pyramids and Cones

Formula for the **surface area** of a **pyramid** or **cone**

$$S = B + L$$

S → Surface Area

B → Base Area → base side $(b)^2$ → $\frac{1}{2} b \times$ base height (h) → π (3.14) radius $(r)^2$

L → Lateral Area → $\frac{1}{2}$ slant height $(\ell) \times$ base perimeter (P) → $\frac{1}{2}$ slant height $(\ell) \times$ base perimeter (P) → π (3.14) radius $(r) \times$ slant height (ℓ)

Square Pyramid

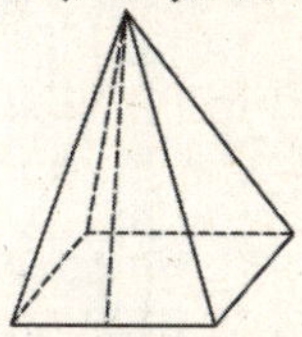

Triangular Pyramid

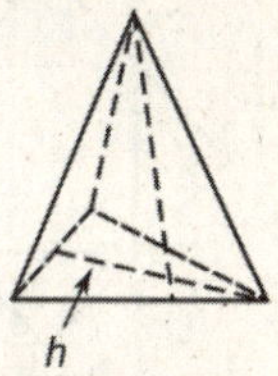

Cone

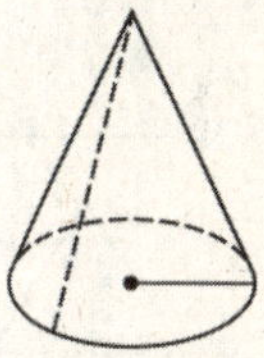

1. For the figure shown, what is:
   (a) the length of *b*? ____
   (b) the length of ℓ? ____
   (c) the surface area? ____

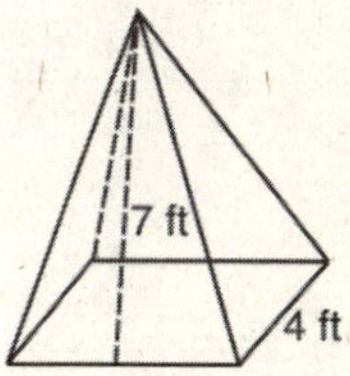

2. For the figure shown, what is:
   (a) the length of *b*? ____
   (b) the length of *h*? ____
   (c) the length of ℓ? ____
   (d) the surface area? ____

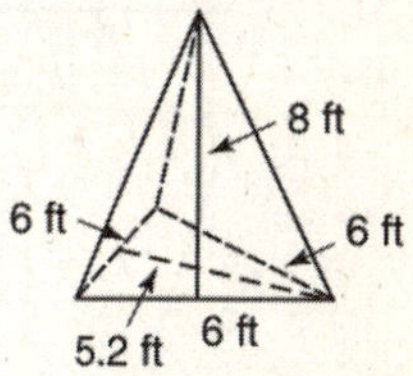

3. For the figure shown, what is:
   (a) the length of *r*? ____
   (b) the length of ℓ? ____
   (c) the surface area? ____

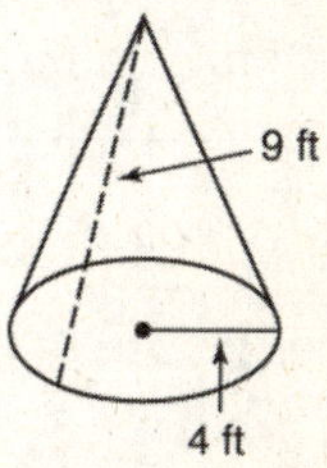

Name ______________________ Date ____________ Class ____________

LESSON 10-5

Puzzles, Twisters & Teasers

Seeing Is Believing

Some of the shapes below are nets that fold to form pyramids. Others do not. Trace each shape, then cut out and fold along the dotted lines. For each net that forms a pyramid, determine the surface area.

1.

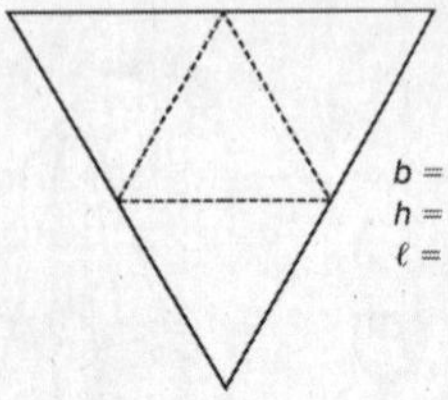

$b = 5$ m
$h = 4.3$ m
$\ell = 5$ m

2.

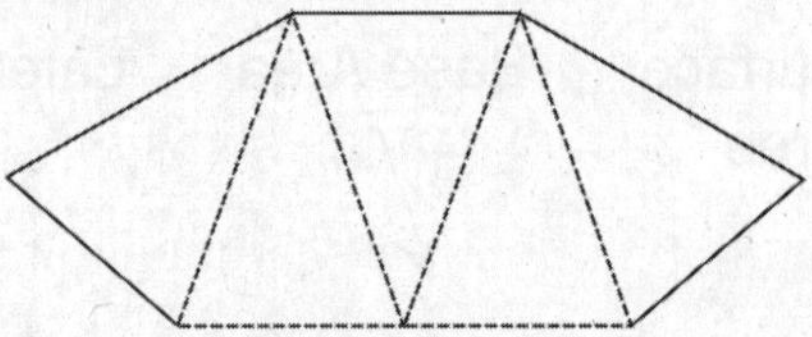

$b = 6$ cm
$\ell = 8$ cm

3.

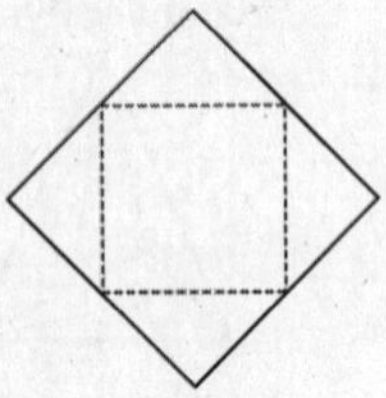

$b = 9$ in.
$\ell = 6$ in.

4.

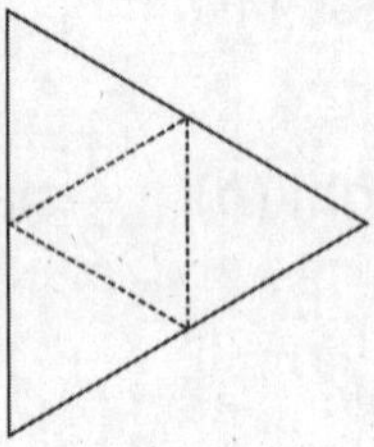

$b = 7$ ft
$h = 6$ ft
$\ell = 7$ ft

5.

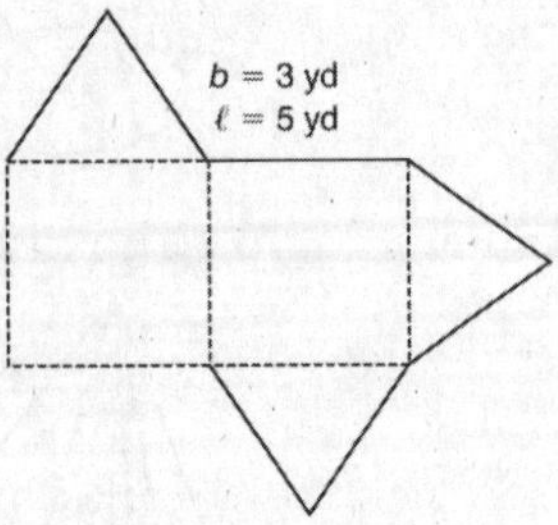

$b = 3$ yd
$\ell = 5$ yd

6.

$b = 8.5$ cm
$\ell = 17$ cm

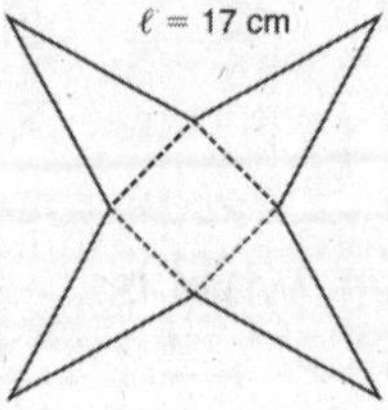

7.

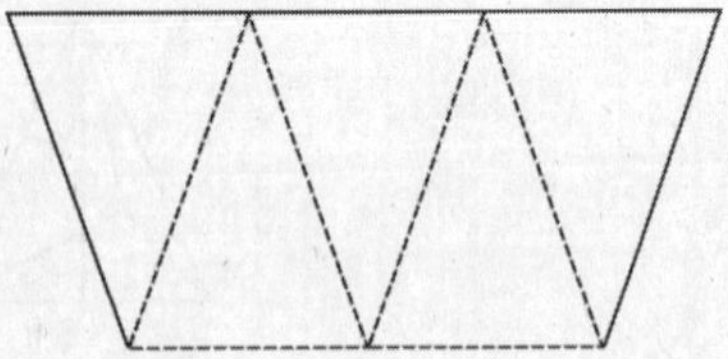

$b = 6$ in.
$\ell = 8$ in.

Name ______________________ Date ______________ Class ______________

LESSON 10-6

Practice A

Changing Dimensions

Two prisms are similar. The scale factor and the surface area of the smaller prism are given in Column A. Draw a line to the surface area of the larger prism in Column B.

Column A Prisms	**Column B Prisms**
1. scale factor: 3 surface area of the smaller prism: 9 cm^2	A 96 cm^2
2. scale factor: 4 surface area of the smaller prism: 5 cm^2	B 90 cm^2
3. scale factor: 5 surface area of the smaller prism: 4 cm^2	C 80 cm^2
4. scale factor: 2 surface area of the smaller prism: 24 cm^2	D 100 cm^2
	E 81 cm^2

Two prisms are similar. The scale factor and the volume of the smaller prism are given in Column A. Draw a line to the volume of the larger prism in Column B.

Column A Prisms	**Column B Prisms**
5. scale factor: 3 volume of the smaller prism: 5 ft^3	F 125 ft^3
6. scale factor: 4 volume of the smaller prism: 2 ft^3	G 128 ft^3
7. scale factor: 5 volume of the smaller prism: 1 ft^3	H 135 ft^3
8. scale factor: 2 volume of the smaller prism: 17 ft^3	I 136 ft^3
	J 144 ft^3

9. A fish tank holds 6 gallons of water. A larger, similarly shaped fish tank has a scale factor of 2. How many times more water does the larger fish tank hold than the smaller fish tank? ______________

Name ______________________ Date ______________ Class ______________

LESSON 10-6

Practice B

Changing Dimensions

Given the scale factor, find the surface area of the similar prism.

1. The scale factor of two similar rectangular prisms is 3. The surface area of the smaller prism is 15 in^2. ______________

2. The scale factor of two similar triangular prisms is 2. The surface area of the smaller prism is 25 cm^2. ______________

3. The scale factor of two similar rectangular prisms is $\frac{1}{2}$. The surface area of the larger prism is 960 ft^2. ______________

4. The scale factor of two similar triangular prisms is 5. The surface area of the smaller prism is 10 m^2. ______________

5. The scale factor of two similar rectangular prisms is $\frac{1}{5}$. The surface area of the larger prism is 625 in^2. ______________

6. The scale factor of two similar pentagonal prisms is 4. The surface area of the smaller prism is 16 cm^2. ______________

Given the scale factor, find the volume of the similar prism.

7. The scale factor of two similar triangular prisms is 3. The volume of the smaller prism is 8 in^3. ______________

8. The scale factor of two similar rectangular prisms is $\frac{1}{2}$. The volume of the larger prism is 648 m^3. ______________

9. The scale factor of two similar triangular prisms is 4. The volume of the smaller prism is 10 cm^3. ______________

10. The scale factor of two similar rectangular prisms is $\frac{1}{4}$. The volume of the larger prism is 1,920 ft^3. ______________

11. The scale factor of two similar triangular prisms is 2. The volume of the smaller prism is 72 yd^3. ______________

12. A small tank weighs 24 pounds when it is full of water. A larger tank that is similar in shape has a scale factor of 3. How much does the larger tank weigh when filled with water? ______________

Name ________________ Date ____________ Class ____________

LESSON 10-6

Practice C

Changing Dimensions

Given the scale factor, find the surface area of the similar prism.

1. The scale factor of two similar rectangular prisms is 5. The surface area of the smaller prism is 13 in^2. ____________
2. The scale factor of two similar triangular prisms is $\frac{1}{8}$. The surface area of the larger prism is 1,024 cm^2. ____________
3. The scale factor of two similar rectangular prisms is 4. The surface area of the smaller prism is 32 ft^2. ____________
4. The scale factor of two similar triangular prisms is $\frac{1}{3}$. The surface area of the larger prism is 828 m^2. ____________
5. The scale factor of two similar rectangular prisms is 3. The surface area of the smaller prism is 13.5 yd^2. ____________

Given the scale factor, find the volume of the similar prism.

6. The scale factor of two similar triangular prisms is 5. The volume of the smaller prism is 12 in^3. ____________
7. The scale factor of two similar rectangular prisms is 3. The volume of the smaller prism is 75 m^3. ____________
8. The scale factor of two similar triangular prisms is $\frac{1}{2}$. The volume of the larger prism is 976 cm^3. ____________
9. The scale factor of two similar rectangular prisms is 5. The volume of the smaller prism is 2.5 ft^3. ____________
10. The scale factor of two similar triangular prisms is $\frac{1}{4}$. The volume of the larger prism is 1,120 cm^3. ____________
11. A large fish tank is made of 2,376 square inches of glass. A smaller, similarly shaped fish tank has a scale factor of $\frac{1}{2}$. Did it take more or less than 1,000 square inches of glass to make the smaller fish tank? Explain your answer.

____

____

____

Name ______________________ Date ______________ Class ______________

LESSON 10-6

Review for Mastery

Changing Dimensions

Changing the dimensions of a three-dimensional figure affects its surface area and its volume.

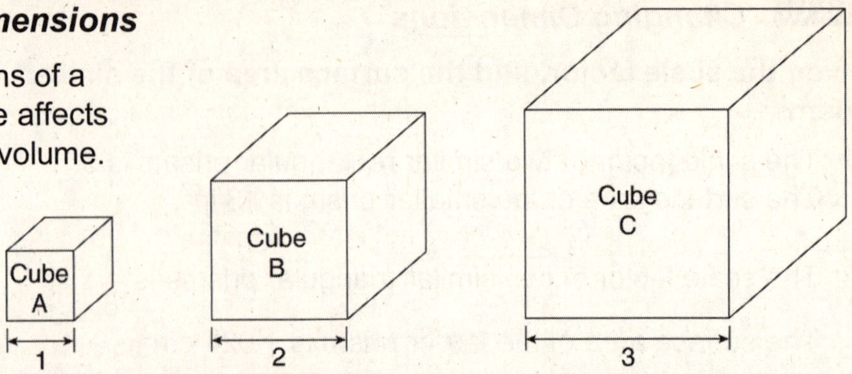

1. Complete the table for the cubes shown above.

Length of Side	Area of One Face	Surface Area	Volume
1	unit^2	units^2	unit^3
2	units^2	units^2	units^3
3	units^2	units^2	units^3

2. The length of one edge of B is ______ times the length of one edge of A.

   The area of one face of B is ______ times the area of one face of A.

   The surface area of B is ______ times the surface area of A: 2^2 = ______.

   The volume of B is ______ times the volume of A: 2^3 = ______.

3. The length of one edge of C is ______ times the length of one edge of A.

   The area of one face of C is ______ times the area of one face of A.

   The surface area of C is ______ times the surface area of A: 3^2 = ______.

   The volume of C is ______ times the volume of A: 3^3 = ______.

For any set of similar three-dimensional figures:

- The surface area of the larger figure equals the surface area of the smaller figure times the scale factor squared.
- The volume of the larger figure equals the volume of the smaller figure times the scale factor cubed.

The scale factor of two similar prisms is 4.

4. The surface area of the smaller prism is 9 in^2. Multiply by ______, or ______.

   The surface area of the larger prism is ______ in^2.

Name ______________________ Date ______________ Class ______________

LESSON 10-6

Challenge

Eggs-actly

The volumes of two similar-shaped figures are proportional to the cube of their lengths.

Suppose an extra-large egg is 2.5 inches long, compared to a smaller egg 1.25 inches long. What is the ratio of their volumes?

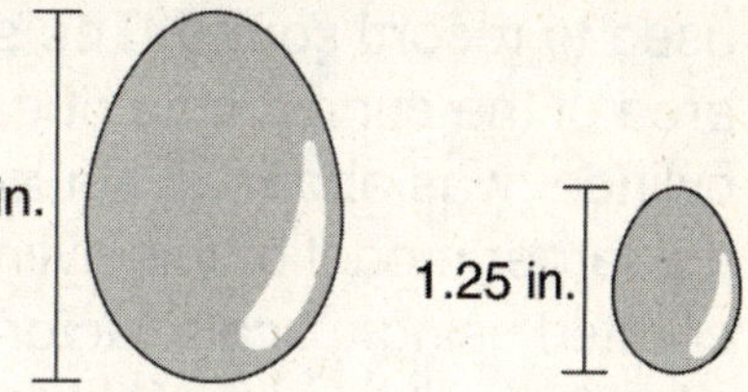

Cube the ratio of their lengths.

$$\left(\frac{2.5}{1.25}\right)^3 = \left(\frac{2}{1}\right)^3 = \frac{8}{1}$$

The ratio of their volumes is 8:1.

So, the larger egg has 8 times as much volume, 8 times as much weight, and 8 times as much food.

Solve. Show your work.

1. One egg is 2 inches long. Another is 2.75 inches long. How many times greater is the volume of the larger egg?

____

2. One egg is 2.25 inches long. Another is 1.5 inches long. How many times greater is the weight of the larger egg?

____

3. The ratio of the lengths of two eggs is 5 to 2. The larger egg contains how many times the food of the smaller egg?

____

4. An egg is 2 inches long. What is the approximate length of an egg with twice the volume?

____

5. An egg is 3 inches long. What is the approximate length of an egg with half the weight?

____

6. An egg is 1.75 inches long. What is the approximate length of an egg with twice the food?

____

Name ______________________ Date ______________ Class ______________

LESSON 10-6

Problem Solving

Changing Dimensions

Write the correct answer.

1. In the late 1800s, wax cylinders were used to record sound. The surface area of the curved side of each cylinder was about 25 square inches. If a larger model of this cylinder is created using a scale factor of 3, what is the lateral surface area of the model?

____

2. A 5-foot wide, 100-foot tall cylindrical water tower was built in St. Louis in the early 1800s. An architecture student wants to build a model using a scale factor of $\frac{1}{6}$. What will be the volume of the model to the nearest cubic foot?

____

3. A cone-shaped plastic cup holds 24 ounces of water. A smaller cup has a scale factor of $\frac{1}{2}$. How much water does the smaller cup hold?

____

4. In the game of Ring Taw, players use a shooting marble that has a surface area of 1.77 square inches. What is the surface area of a large ball if the scale factor is 5?

____

Choose the letter of the correct answer.

5. The volume of a rectangular prism is 48 cubic centimeters. The volume of a similar rectangular prism is 6 cubic centimeters What is the scale factor for the rectangles?

   A 8   C 4

   B 6   D 2

6. A cooking pot used in the cafeteria weighs 64 pounds when it is filled with soup. How much would a similar pot with a scale factor of $\frac{1}{2}$ weigh when filled with the same soup?

   F 4 lb   H 16 lb

   G 8 lb   J 32 lb

7. For his science project, Marty is building a model of Pluto, which has a surface area of about 6,376,000 square miles. He plans to cover his model with red foil. If he uses a scale factor of 5,000, how much red foil will he need to the nearest hundredth of a square mile?

   A 1.26 mi^2 C 0.26 mi^2

   B 0.026 mi^2 D 0.126 mi^2

8. The scale factor of two similar triangular prisms is 5. What is the possible surface area of both prisms?

   F 2,000 cm^2 and 80 cm^2

   G 125 cm^2 and 25 cm^2

   H 5,000 cm^2 and 40 cm^2

   J 10,000 cm^2 and 60 cm^2

Name ______________________ Date ______________ Class ______________

LESSON 10-6

Reading Strategies

Organize Information

Each face on this cube measures 1 unit by 1 unit. The area of one face is 1 square unit.

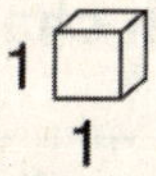

Adding 6 faces = 6 square units. The surface area is 6 $units^2$.

Each face on this cube measures 2 units by 2 units.

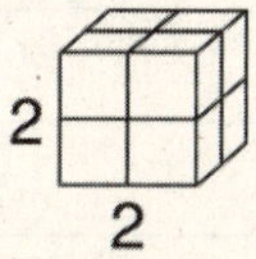

1. What is the area of one face? ______________
2. What is the total area of all six faces? ______________

You can use a table to organize the information about the two cubes.

Dimensions of Each Face	Area of 1 Face	Total Surface Area
1 unit by 1 unit	1 $unit^2$	$6 \cdot 1 = 6$ $units^2$
2 units by 2 units	4 $units^2$	$6 \cdot 4 = 24$ $units^2$

This cube would be the next cube in the list.

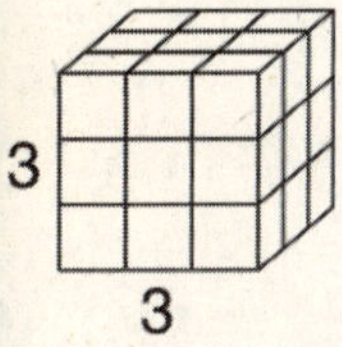

Use this cube to answer the following questions.

3. What are the dimensions of each face on the cube? ______________
4. What is the area of one face of the cube?

____

5. How many faces does the cube have? ______________
6. What is the surface area for the cube?

____

7. If the list above continued, what would be the dimensions of each face on the next cube? ______________

Name ______________________ Date ______________ Class ______________

LESSON 10-6

Puzzles, Twisters & Teasers

A New Dimension!

Complete both charts. Use your answers to solve the riddle.

scale factor	smaller surface area	larger surface area
4	35 in^2	**T**
6	24 m^2	**S**
8	40 ft^2	**U**
9	**H**	1,620 m^2

scale factor	smaller volume	larger volume
2	29 ft^3	**C**
3	18 lb^3	**E**
5	8 cm^3	**R**
4	17 in^3	**O**
4	32 lb^3	**A**
2	**K**	600 in^3

Why did the hairdresser win the race?

___ ___ ___   ___ ___ ___ ___   ___
864 20 486   560 1,088 1,088 75   2,048

___ ___ ___ ___ ___ ___ ___ ___.
864 20 1,088 1,000 560 232 2,560 560

Answers

LESSON 10-1

Practice A

1. triangle, triangular pyramid
2. rectangle, rectangular prism
3. square, square pyramid
4. hexagon, hexagonal prism
5. triangle, triangular prism
6. pentagon, pentagonal pyramid
7. not a polyhedron; cone
8. polyhedron; octagonal prism
9. not a polyhedron; sphere

Practice B

1. one square base; four triangular faces; square pyramid
2. one triangular base; three triangular faces; triangular pyramid
3. two pentagonal bases; five rectangular faces; pentagonal prism
4. two rectangular bases; four rectangular faces; rectangular prism
5. one hexagonal base; six triangular faces; hexagonal pyramid
6. two octagonal bases; eight rectangular faces; octagonal prism
7. polyhedron; hexagonal prism
8. not a polyhedron; cylinder
9. not a polyhedron; sphere
10. polyhedron; triangular prism
11. not a polyhedron; cone
12. polyhedron; pentagonal pyramid

Practice C

1. one rectangular base; four triangular faces; rectangular pyramid
2. two triangular bases; three rectangular faces; triangular prism
3. two octagonal bases; eight rectangular faces; octagonal prism
4. polyhedron; pentagonal pyramid
5. not a polyhedron; cone
6. polyhedron; hexagonal prism
7. rectangular prism
8. square pyramid
9. triangular pyramid
10. pentagonal prism
11. cylinder
12. hexagonal pyramid

Review for Mastery

1. square; triangles
2. pentagons; rectangles
3. yes
4. no
5. yes

Challenge

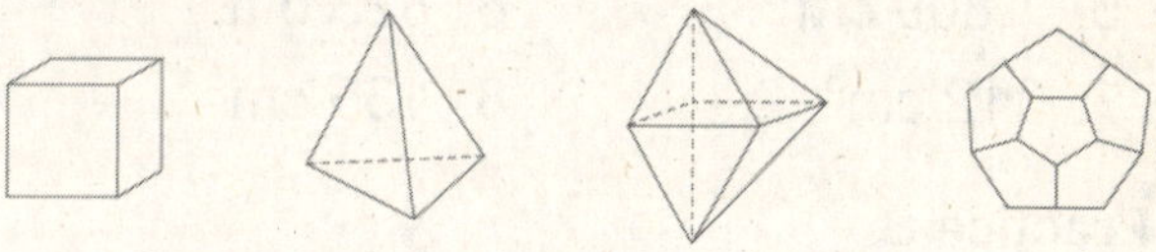

1. 6 faces; square prism or cube
2. 4 faces; triangular pyramid
3. 8 faces
4. 12 faces

Problem Solving

1. pyramid; The top of the tower is a vertex, and the sides are triangular.
2. a prism; It has triangular bases and parallelograms form the other faces.
3. a cone
4. a cylinder
5. B
6. H
7. A

Reading Strategies

1. polyhedron
2. a face
3. a prism
4. a pyramid
5. bases
6. the base
7. The surfaces of a cone are not all polygons.
8. Possible answer: cylinder

Puzzles, Twisters and Teasers

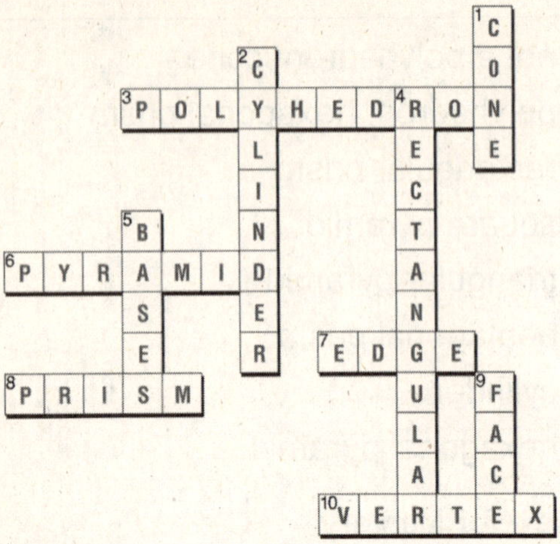

LESSON 10-2

Practice A

1. D
2. H
3. C
4. G
5. 1,809 cm^3
6. 823.6 ft^3
7. 312 cm^3
8. 135 cm^3

Practice B

1. 180 in^3
2. 225 yd^3
3. 57.498 in^3
4. 729 in^3
5. 357.5 cm^3
6. 1,053 m^3
7. 953.8 cm^3
8. 1,538 in^3
9. 2,984.3 m^3
10. 3,886 cm^3

Practice C

1. 124.4 in^3
2. 477.8 cm^3
3. 335.8 m^3
4. 53.9 ft^3
5. 13.0 m^3
6. 20.2 cm^3
7. 176.6 in^3
8. 12 cm
9. 240 in^3
10. 1,240.5 cm^3
11. 548.9 yd^3
12. 780.7 ft^3

Review for Mastery

1. 6
2. 3
3. 18
4. triangle
5. 6; 3; 9
6. 5
7. 9; 5; 45
8. circle
9. 4; 50
10. 3
11. 50; 3; 150

Challenge

1. 8 cubes
2. 8 cubes; 0 cubes; 0 cubes; 0 cubes
3. 27 cubes
4. 8 cubes; 12 cubes; 6 cubes; 1 cube
5. 64 cubes
6. 8 cubes; 24 cubes; 24 cubes; 8 cubes
7.

Painted on 3 Faces	Painted on 2 Faces	Painted on 1 Face	Painted on 0 Faces
8	0	0	0
8	12	6	1
8	24	24	8
8	36	54	27

8. always 8

Problem Solving

1. 3,768 ft^3
2. 24.2 L
3. 1.2 gal
4. 2.7 in^3
5. C
6. J
7. C
8. G

Reading Strategies

1. 3 units
2. 2 units
3. 3 units
4. 18 cubic units
5. 4 units; 2 units
6. 3 units
7. 4 • 2 • 3 = 24 cubic units

Puzzles, Twisters & Teasers

1. P
2. Y
3. U
4. O
5. E
6. C
7. M
8. R
9. K

Y O U C R A C K
M E U P

LESSON 10-3

Practice A

1. $V = \frac{1}{3}Bh$
2. $V = \frac{1}{3}\pi r^2 h$
3. 36 m^3
4. 500 ft^3
5. 6 cm^3
6. 38 yd^3
7. 8 in^3
8. 2,826 mm^3
9. 540 m^3
10. 33 in^3
11. 360 ft^3

Practice B

1. 333.7 yd^3
2. 116.7 in^3
3. 108 cm^3
4. 35 m^3
5. 200 ft^3
6. 105 m^3
7. 201.0 ft^3
8. 198.9 m^3
9. 32.7 in^3
10. 39.3 cm^3
11. 92.3 ft^3
12. 147.7 in^3

Practice C

1. 60.2 m^3
2. 600 ft^3
3. 49.1 cm^3
4. 410.3 yd^3
5. 398.0 in^3
6. 14.8 m^3
7. 63 ft^3
8. 736.9 m^3
9. 37.3 ft^3
10. 1,177.5 in^3
11. 190.0 in^3
12. 20 cm^3

Review for Mastery

1. 5; 3; 15
2. 6
3. 15; 6; 30
4. 42 cm^3
5. 22 yd^3
6. 96 in^3
7. 9 m^3
8. 2; 12.56
9. 12.56; 5; 62.8; 21
10. 5; 25; 78.5; 78.5; 4 314; 105
11. 75 cm^3
12. 733 ft^3
13. 339 cm^3
14. 469 ft^3

Challenge

1. 216 m^3
24 m^3
192 m^3
2. 640 ft^3
213 ft^3
427 ft^3
3. 216 m^3
96 m^3
120 m^3
4. 960 cm^3
240 cm^3
720 cm^3

Problem Solving

1. 94.2 mL
2. 90,428,160 ft^3
3. 392.5 cm^3
4. The figures have the same volume. $V(\text{prism}) = Bh$; $V(\text{pyramid}) = (\frac{1}{3}Bh) \cdot 3 = Bh$
5. A
6. J
7. A
8. J

Reading Strategies

1. a cone
2. a polygon
3. a circle
4. πr^2
5. triangle
6. In each you find $\frac{1}{3}$ of the area of the base multiplied by the height.

Puzzles, Twisters & Teasers

C	O	N	G	R	U	E	N	T	C	V
L	U	J	I	A	C	V	B	N	O	W
O	O	P	L	D	F	G	H	J	N	C
U	D	P	R	I	S	M	E	R	T	Y
D	M	N	B	U	A	S	D	F	A	L
P	C	B	A	S	E	T	Y	U	I	I
Y	W	D	R	T	H	U	I	J	N	N
R	W	S	C	D	E	C	O	N	E	D
A	T	H	U	Y	S	P	H	E	R	E
M	V	O	L	U	M	E	G	T	R	R
I	P	L	M	N	K	O	I	J	B	Y
D	Q	W	E	R	T	Y	U	I	I	O

cloud

LESSON 10-4

Practice A

1. D
2. H
3. C
4. F
5. 407 in^2

Practice B

1. 142 in^2
2. 600 mm^2
3. 37.7 in^2
4. 678.2 in^2
5. 165.1 ft^2

Practice C

1. 102.1 in^2
2. 217.7 cm^2
3. 525.3 cm^2
4. 302.1 ft^2
5. 176.4 yd^2

Review for Mastery

1. 8; 5; 40;
   40; 80
2. 5; 3; 15;
   15; 30
3. 8; 3; 24;
   24; 48
4. 80; 30; 48;
   158; 158
5. 340 in^2
6. 592 cm^2
7. 5^2; 78.5
8. 78.5; 157
9. 5; 8; 251.2
10. 157; 251.2; 408
11. 603 cm^2
12. 1,130 $in.^2$

Challenge

1.

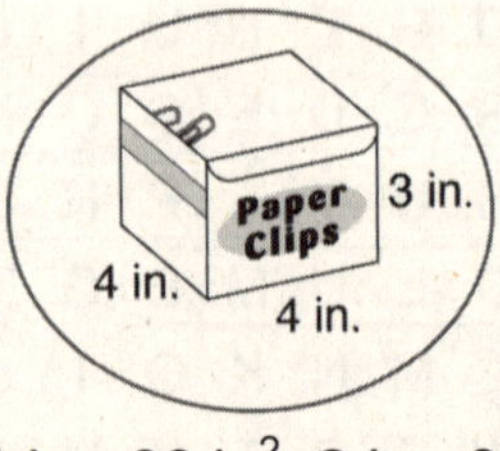

80 in^2; 4 in.; 88 in^2; 2 in.; 92 in^2

2.

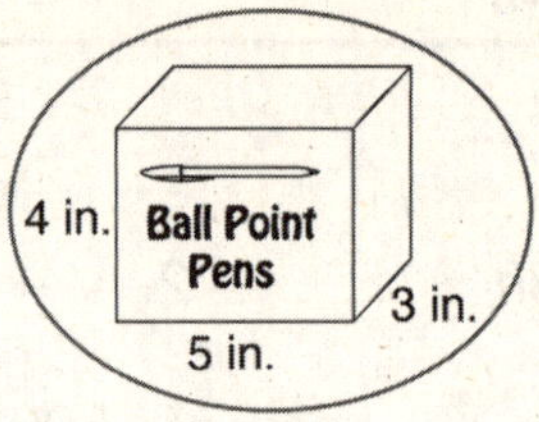

94 in^2; 10 in.; 112 in^2; 6 in.; 104 in^2

3.

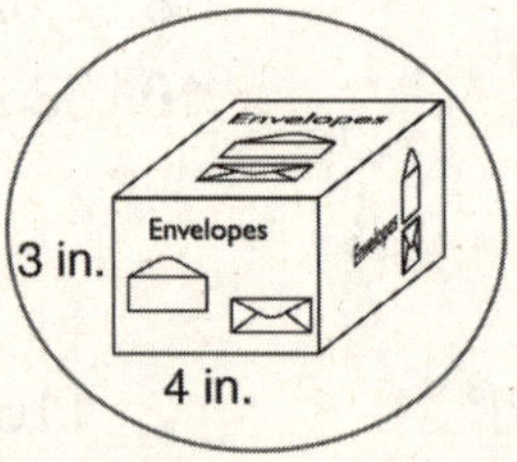

4 in.; 124 in^2; 6 in.; 108 in^2;
3 in.; 114 in^2

Problem Solving

1. 42.39 in^2
2. 872 in^2
3. 1,884 cm^2
4. 32 ft^2
5. B
6. H
7. C
8. G

Reading Strategies

1. 48 in^2
2. 48 in^2
3. 32 in^2
4. 32 in^2
5. 24 in^2
6. 24 in^2
7. 208 in^2

Puzzles, Twisters & Teasers

L 256.2 ft^2
B 244.9 m^2
A 562 m^2
H 653.1 in^2
T 286 ft^2
O 351.7 ft^2
S 1,160 ft^2
I 268 m^2

IT WAS A MOTHBALL

LESSON 10-5

Practice A

1. 20 in^2
2. 72 yd^2
3. 85 cm^2
4. 156 ft^2
5. 43 ft^2
6. 78.6 in^2
7. 136 yd^2
8. 85.75 m^2
9. 141.3 m^2
10. 138.16 cm^2
11. 207.24 ft^2
12. 197.82 yd^2

Practice B

1. 68 cm^2
2. 129.25 ft^2
3. 299.25 ft^2
4. 480 in^2
5. 105 yd^2
6. 150.75 m^2
7. 170.1 m^2
8. 177 cm^2
9. 138.2 ft^2
10. 141.3 yd^2
11. 207.24 $in.^2$
12. 197.82 cm^2

Practice C

1. 134.75 cm^2
2. 272.16 ft^2
3. 722.25 ft^2
4. 663 in^2
5. 178 yd^2
6. 256.75 in^2
7. 342.72 m^2
8. 462.21 cm^2
9. 678.24 ft^2
10. 1,318.8 m^2
11. 197.82 $in.^2$
12. 728.48 m^2

Review for Mastery

1. 5 m
2. Figures A and C
3. The perimeter of Figure A is 16 m. The perimeter of Figure B is 15 m.
4. 78.5 m^2

Challenge

1. 13 cm
2. 5 cm
3. You can derive the formula by starting with the formula for the surface area of a cone. Since the surface area of a cone is $\pi r^2 + \pi r\ell$ and since πr^2 is the area of the circular base, then the area of the cone excluding the base is $\pi r\ell$. You can substitute for ℓ using the formula $\ell = \sqrt{r^2 + H^2}$. The formula for the surface area of a cone without its base is $\pi r \sqrt{r^2 + H^2}$.
4. Possible answer: the amount of material needed to construct a tepee, which uses the ground as a base
5. They are essentially the same relationship. The altitude and the base of the cone form the legs of a right triangle whose hypotenuse is the slant height of the cone. Therefore, $\ell = \sqrt{r^2 + H^2}$ and $\sqrt{a^2 + b^2} = c$ are equivalent.

Problem Solving

1. 11.25 cm
2. $r = 20$ and $\ell = 50$
3. The surface area is about 99 m^2.
4. 1,397.5 square meters
5. C
6. A

Reading Strategies

1. (a) 4 ft
   (b) 7 ft
   (c) 72 ft^2
2. (a) 6 ft
   (b) 5.2 ft
   (c) 8 ft
   (d) 87.6 ft^2
3. (a) 4 ft
   (b) 9 ft
   (c) 163.28 ft^2

Puzzles, Twisters & Teasers

1. 43 m^2
2. Not a pyramid
3. 189 $in.^2$
4. 84 ft^2
5. Not a pyramid
6. 361.25 cm^2
7. Not a pyramid

LESSON 10-6

Practice A

Column A Prisms

1. scale factor: 3 surface area of the smaller prism: 9 cm^2
2. scale factor: 4 surface area of the smaller prism: 5 cm^2
3. scale factor: 5 surface area of the smaller prism: 4 cm^2
4. scale factor: 2 surface area of the smaller prism: 24 cm^2
5. scale factor: 3 volume of the smaller prism: 5 ft^3
6. scale factor: 4 volume of the smaller prism: 2 ft^3
7. scale factor: 5 volume of the smaller prism: 1 ft^3
8. scale factor: 2 volume of the smaller prism: 17 ft^3

Column B Prisms

A 96 cm^2
B 90 cm^2
C 80 cm^2
D 100 cm^2
E 81 cm^2
F 125 ft^3
G 128 ft^3
H 135 ft^3
I 136 ft^3
J 144 ft^3

9. 8 times more

Practice B

1. 135 in^2
2. 100 cm^2
3. 240 ft^2
4. 250 m^2
5. 25 in^2
6. 256 cm^2
7. 216 in^3
8. 81 m^3
9. 640 cm^3
10. 30 ft^3
11. 576 yd^3
12. 648 lb

Practice C

1. 325 in^2
2. 16 cm^2
3. 512 ft^2
4. 92 m^2
5. 121.5 yd^2
6. 1,500 in^3
7. 2,025 m^3
8. 122 cm^3
9. 312.5 ft^3
10. 17.5 cm^3
11. It took less than 1,000 square inches. Since the scale factor is $\frac{1}{2}$, the surface area of the smaller tank is $\frac{1}{2} \bullet \frac{1}{2}$, or $\frac{1}{4}$ as great as 2,376. $\frac{1}{4} \bullet 2.376 = 594$, and $594 < 1{,}000$.

Review for Mastery

1.

Area of One Face	Surface Area	Volume
1 $unit^2$	6 $units^2$	1 $unit^3$
4 $units^2$	24 $units^2$	8 $units^3$
9 $units^2$	54 $units^2$	27 $units^3$

2. 2;
 4;
 4; 4;
 8; 8
3. 3;
 9;
 9; 9;
 27; 27
4. 4^2; 16;
 144

Challenge

1. $\left(\frac{2.75}{2}\right)^3 \approx \frac{20.8}{8} = 2.6$; about 2.6 times

2. $\left(\frac{2.25}{1.5}\right)^3 \approx \left(\frac{3}{2}\right)^3$; about 3.4 times

3. $\left(\frac{5}{2}\right)^3 \approx \frac{125}{8} = 15.625$; about 15.6 times as much

4. about 2.5 in.

5. about 2.4 in.

6. about 2.2 in.

Problem Solving

1. about 225 in^2
2. 9 ft^3
3. 3 oz
4. 44.25 in^2
5. D
6. G
7. C
8. F

Reading Strategies

1. 4 $units^2$
2. 24 $units^2$
3. 3 units by 3 units
4. 3 units • 3 units = 9 square units or 9 $units^2$
5. 6
6. 6 units • 9 units = 54 $units^2$
7. 4 units by 4 units

Puzzles, Twisters & Teasers

scale factor	smaller surface area	larger surface area
4	35 in^2	560 in^2 **T**
6	24 m^2	864 m^2 **S**
8	40 ft^2	2,560 ft^2 **U**
9	20 m^2 **H**	1,620 m^2

scale factor	smaller volume	larger volume
2	29 ft^3	232 ft^3 **C**
3	18 lb^3	486 lb^3 **E**
5	8 cm^3	1,000 cm^3 **R**
4	17 in^3	1,088 in^3 **O**
4	32 lb^3	2,048 lb^3 **A**
2	75 in^3 **K**	600 in^3

S H E T O O K A S H O R T C U T